More Praise for

LIGHTS OUT

"A bold enumeration of the challenges posed by the digital age; an appeal to safeguard new instruments of human flourishing by studying the ways in which they could be exploited."

—*Henry A. Kissinger*

"Try to imagine what a malevolent government, armed with the latest computer sophistication, could do to another nation's complex and entirely digital-dependent economy and social infrastructure. Fortunately, Ted Koppel has imagined it for us. We have been warned."

—*George F. Will*

"In *Lights Out,* Ted Koppel uses his profound journalistic talents to raise pressing questions about our nation's aging electrical grid. Through interview after interview with leading experts, Koppel paints a compelling picture of the impact cyberattacks may have on the grid. The book reveals the vulnerability of perhaps the most critical of all the infrastructures of our modern society: the electricity that keeps our modern society humming along."

—*Marc Goodman, author of* Future Crimes

"When the lights go out after the cyberattack, this is the book everyone will read."

—*Richard A. Clarke, author of* Cyber War *and former National Coordinator for Security, Infrastructure Protection, and Counter-terrorism*

LIGHTS OUT

A Cyberattack

A Nation Unprepared

Surviving the Aftermath

TED KOPPEL

CROWN PUBLISHERS
NEW YORK

Copyright © 2015 by Edward J. Koppel

All rights reserved.
Published in the United States by Crown Publishers, an imprint of the
Crown Publishing Group, a division of Penguin Random House LLC,
New York.
www.crownpublishing.com

CROWN and the Crown colophon are registered trademarks of
Penguin Random House LLC.

Library of Congress Cataloging-in-Publication Data is available upon
request.

ISBN 978-0-553-41996-2
eBook ISBN 978-0-553-41997-9

Printed in the United States of America

Jacket design by Tal Goretsky
Jacket photograph by NOAA/Science Source

10 9 8 7 6 5 4 3 2 1

First Edition

To our grandchildren:

Jake and Dylan, Aidan, Alice and Annabelle, Cole and

Grace Ann(e). Here's hoping that Opi got it wrong.

CONTENTS

....

PART I: A CYBERATTACK

PART II: A NATION UNPREPARED

CONTENTS

PART III: SURVIVING THE AFTERMATH

Part I

A CYBERATTACK

1

....

Warfare 2.0

Everyone is not entitled to his own facts.
—DANIEL PATRICK MOYNIHAN

Darkness.

Extended periods of darkness, longer and more profound than anyone now living in one of America's great cities has ever known.

As power shuts down there is darkness and the sudden loss of electrical conveniences. As batteries lose power, there is the more gradual failure of cellphones, portable radios, and flashlights.

Emergency generators provide pockets of light and power, but there is little running water anywhere. In cities with water towers on the roofs of high-rise buildings, gravity keeps the flow going for two, perhaps three days. When this runs out, taps go dry; toilets no longer flush. Emergency supplies of bottled water are too scarce to use for anything but drinking, and there is nowhere to replenish the supply. Disposal of human waste becomes a critical issue within days.

Supermarket and pharmacy shelves are empty in a matter of hours. It is a shock to discover how quickly a city can exhaust its food supplies. Stores do not readily adapt to panic buying, and many city dwellers, accustomed to ordering out, have only scant supplies at home. There is no immediate resupply, and people become desperate.

For the first couple of days, emergency personnel are overwhelmingly engaged in rescuing people trapped in elevators. Medicines are running out. Home care patients reliant on ventilators and other medical machines are already dying. One city has hoisted balloons marking the sites of generators, hauled out of storage to serve new emergency centers. Almost everyone needs some kind of assistance, and no one has adequate information.

The city has flooded the streets with police to preserve calm, to maintain order, but the police themselves lack critical information. People are less concerned with what exactly happened than with how long it will take to restore power. This is a society that regards information, the ability to communicate instantly, as an entitlement. Round-the-clock chatter on radio and television continues, but there's little new information and a diminishing number of people still have access to functioning radios and television sets. The constant barrage of messages that once flowed between iPhones and among laptops has sputtered to a trickle. The tissue of emails, texts, and phone calls that held our social networks together is tearing.

There is a growing awareness that this power outage extends far beyond any particular city and its suburbs. It may extend over several states. Tens of millions of people appear affected. Fuel is beginning to run out. Operating gas stations have no way of determining when their supply of gasoline and diesel will be replenished, and gas stations without backup generators are unable to operate their pumps. Those with generators are running out of fuel and shutting down.

The amount of water, food, and fuel consumed by a city of several million inhabitants is staggering. Emergency supplies are sufficient only for a matter of days, and official estimates of how much help is needed and how soon it can be delivered are vague, uncertain. The majority who believed that power outages are limited in duration, that help always arrives from beyond the edge of darkness, is undergoing a crisis of conviction. The assumption that the city, the state, or even the federal government has the plans and the wherewithal to handle this particular crisis is being replaced by the terrible sense that people are increasingly on their own. When that awareness takes hold it leads to a contagion of panic and chaos.

There are emergency preparedness plans in place for earthquakes and hurricanes, heat waves and ice storms. There are plans for power outages of a few days, affecting as many as several million people. But if a highly populated area was without electricity for a period of months or even weeks, there is no master plan for the civilian population.

Preparing for doomsday has its own rich history in this country, and predictions of the apocalypse are hardly new to people of my generation. We lived for decades with the assumption that nuclear war with the Soviet Union was a real possibility. We learned some useful lessons. (We'll ramble through the age of bomb shelters and civil defense in a later chapter.) Ultimately, Moscow and Washington came to the conclusion that mutual assured destruction, holding each other hostage to the fear of nuclear reprisal, was a healthier approach to coexistence than mass evacuation or hunkering down in our respective warrens of bomb shelters in the hopes of surviving a nuclear winter.

We are living in different times. Whether the threat of nuclear war has actually receded or whether we've simply become inured to a condition we cannot change, most of us have finally learned "to stop worrying and love the bomb." In reality, though, the ranks of our enemies, those who would and can inflict serious damage on America, have grown and diversified. So many of our transactions are now conducted in cyberspace that we have developed dependencies we could not even have imagined a generation ago. To be dependent is to be vulnerable. We have grown cheerfully dependent on the benefits of our online transactions, even as we observe the growth of cyber crime. We remain largely oblivious to the potential catastrophe of a well-targeted cyberattack.

On one level, cyber crime is now so commonplace that we have already absorbed it into the catalogue of daily outrages that we observe, briefly register, and ultimately ignore. Over the course of less than a generation, cyber criminals have become adept at using the Internet for robbery on an almost unimaginable scale. Still, despite the media attention generated by the more dazzling smash-and-grab operations, the cyber criminals whose only intention is to siphon off wealth or hijack several million credit card identities should have a lower priority among our concerns. Their goal is merely grand larceny.

More worrisome is the increasing number of cyberattacks designed to vacuum up enormous quantities of data in what appear to be wholesale intelligence gathering operations. The most ambitious of these was announced on June 4, 2015, and targeted the Office of Personnel Management, which handles government security clearances and federal employee records. The *New York Times* quoted J. David Cox Sr., the president of the American Federation of Government Employees, as saying the breach might have affected "all 2.1 million current federal employees and an additional two million federal retirees and former employees." FBI director James Comey told a Senate hearing that the actual number of hacked files was likely more than ten times that number—22.1 million. Government sources were quoted as claiming that the intrusion originated in China. The *Times* report raises a number of relevant issues: The probe was initiated at the end of 2014. It wasn't discovered until April of 2015. It is believed to have originated in China, but the Chinese government denied the charge, challenging U.S. authorities to provide evidence. Producing evidence would reveal highly classified sources and methods. "The most sophisticated attacks," the *Times* noted, "often look as if they were initiated inside the United States, and tracking their true paths can lead down many blind paths." All of these issues will receive further attention in later chapters. But as disturbing as these massive data collection operations may be, even they do not come close to representing the greatest cyber threat. Our attention needs to be focused on those who intend widespread destruction.

The Internet provides instant, often anonymous access to the operations that enable our critical infrastructure systems to function safely and efficiently. In early March 2015 the Government Accountability Office issued a report warning that the air traffic control system is vulnerable to cyberattack. This, the report concluded with commendable understatement, "could disrupt air traffic control operations." Our rail system, our communications networks, and our healthcare system are similarly vulnerable. If, however, an adversary of this country has as its goal inflicting maximum damage and pain on the largest number of Americans, there may not be a more productive target than one of our electric power grids.

Electricity is what keeps our society tethered to modern times. There are three power grids that generate and distribute electricity throughout the United States, and taking down all or any part of a grid would scatter millions of Americans in a desperate search for light, while those unable to travel would tumble back into something approximating the mid-nineteenth century. The very structure that keeps electricity flowing throughout the United States depends absolutely on computerized systems designed to maintain perfect balance between supply and demand. Maintaining that balance is not an accounting measure, it is an operational imperative. The point needs to be restated: for the grid to remain fully operational, the supply and demand of electricity have to be kept in perfect balance. It is the Internet that provides the instant access to the computerized systems that maintain that equilibrium. If a sophisticated hacker gained access to one of those systems and succeeded in throwing that precarious balance out of kilter, the consequences would be devastating. We can take limited comfort in the knowledge that such an attack would require painstaking preparation and a highly sophisticated understanding of how the system works and where its vulnerabilities lie. Less reassuring is the knowledge that several nations already have that expertise, and—even more unsettling—that criminal and terrorist organizations are in the process of acquiring it. Our media report daily on increasingly bold and costly acts of online piracy that are already costing the U.S. economy countless billions of dollars a year. Cyberattacks as instruments of national policy, though, tend to be less visible because neither the target nor the attacker is inclined to publicize the event.

History often provides a lens through which irony comes into focus. The United States, for example, was the first and only nation to have used an atomic weapon, and it has spent the intervening decades trying to limit nuclear proliferation. And the United States, in collaboration with Israel, mounted a hugely successful cyberattack on Iran's nuclear program in 2008 and now finds itself dealing with the consequences of having been the first to use a digital weapon as an instrument of policy. Iran wasted little time in launching what appeared to be a retaliatory cyberattack, choosing to target Aramco in Saudi Arabia, destroying thirty thousand of its computers. Why the Saudi oil giant instead of an American or Israeli target? We can only speculate. Iran may have wanted to issue a warning, demonstrating some of its own cyber capabilities without directly engaging the more dangerous Americans or Israelis. In any event, Iran made its point, and a new style of warfare has, within a matter of only a few years, become commonplace. Russia, China, and Iran, among others, continue on an almost daily basis to demonstrate a range of cyber capabilities in espionage, denial-of-service attacks, and the planting of digital time bombs, capable of inflicting widespread damage on a U.S. power grid or other piece of critical infrastructure.

For several reasons, the clear logic of a swift attack and response that enables a policy of deterrence between nuclear rivals does not yet exist in the world of cyber warfare. For one, cyberattacks can be launched or activated from anywhere in the world. The point at which a command originates is often deliberately disguised so that its electronic instruction appears to be coming from a point several iterations removed from its actual location. It is difficult to retaliate against an aggressor with no return address. Nation-states may be inhibited by the prospect of ultimately being unmasked, but it is not easily or instantly accomplished. For another, the list of capable cyberattackers is far more numerous than the current list of the world's nuclear powers. We literally have no count of how many groups or even individuals are capable of launching truly damaging attacks on our electric power grids—some, perhaps even most of them, uninhibited by the threat of retaliation.

There is scant consolation to be found in the fact that a major attack on the grid hasn't happened yet. Modified attacks on government, banking, commercial, and infrastructure targets are already occurring daily, and while sufficient motive to take out an electric power grid may be lacking for the moment, capability is not. As the ranks of capable actors grow, the bar for cyber aggression is lowered. The unintended consequences of Internet dependency are already piling up. Prudence suggests that we at least consider the possibility of a cyberattack against the grid, the consequences of which would be so devastating that no administration could consider it anything less than an act of war.

This book is about dealing with the consequences of losing power in more than one sense of the word. Without ready access to electricity, we are thrust back into another age—an age in which many of us would lack both the experience and the resources to survive. Precisely how that happens is, ultimately, less important than how prepared we are for the consequences. It would be reassuring to report that the grid is adequately defended against cyberattack. It is not. The grid is a network connecting thousands of companies, many of which still put profit ahead of security. Critical equipment that is decades old and difficult to replace sits in exposed locations, vulnerable to physical attack. Computerized systems that control the flow of electricity around the country were designed before anyone even contemplated cyberspace as an environment suited to malicious attacks. It would be comforting to report that those agencies charged with responding to disaster are adequately prepared to deal with the consequences of a cyberattack on the grid. They are not. The Department of Homeland Security has no plans beyond those designed to deal with the aftermath of natural disasters. The deputy administrator of the Federal Emergency Management Agency (FEMA) believes that a major urban center would have to be evacuated. His boss, the administrator, does not. The administrator believes that a successful cyberattack on a power grid is possible, even likely. His deputy does not. The current secretary of homeland security is sure that a plan to deal with the aftermath of a cyberattack on the grid exists, but he doesn't know any details of the plan. As of this writing, there is no specific plan.

We are unprepared, but why isn't the issue higher on our list of national priorities? It is difficult for anyone holding public office to focus attention on a problem without being able to offer any solutions. Then, too, the American public needs to be convinced that the threat is real. And let the record show: it is not easy to convince the American public of anything.

For example, in February 2014 the British market research company YouGov posed the following question about President Obama to a sampling of one thousand American adults: "Would you say you know for sure he was born outside the U.S., or do you think it is possible he was born in the United States?" Thirty-nine percent of those sampled stated with confidence that they "know for sure he was born outside the U.S." And in 2013 the Pew Research Center released the results of a global survey conducted in thirty-nine countries and involving 37,653 respondents. One of the key questions sought to establish whether people consider global warming a threat to their country. Only 40 percent of Americans did, placing the United States among the least concerned countries, along with China, Pakistan, Egypt, the Czech Republic, Israel, and Jordan. One year later, another Pew Research Center survey examined the impact of political ideology on the issue of global warming. The survey found that 91 percent of people identifying themselves as "solid liberals" believed that "the earth is getting warmer," while only 21 percent of those identifying themselves as "steadfast conservatives" agreed.

Fifty years ago, in the era of Walter Cronkite, Chet Huntley, David Brinkley, Howard K. Smith, and Eric Sevareid, Americans communed by the tens of millions before their television sets in a willing, if temporary, suspension of partisanship. In hindsight, the notion that Cronkite, a television anchorman, polled regularly as "the most trusted man in America" seems quaint. Brinkley twinkled a little, but for the most part, all of those somber, gray men took themselves and their nightly presentation of important events seriously, and their viewing public, without easy alternatives at hand, tended to reflect that broad-based, middle-of-the-road sobriety. There was, even back then, an undercurrent of irritation at the "liberal tendencies" of the television news networks, but this was in an era before CNN, Fox News, and MSNBC—indeed, before the existence of cable television. The broadcast networks were the only game in town. If there was anything approaching a common denominator, providing the American public with a foundation of shared civic values, it was those three newscasts on CBS, NBC, and ABC.

Today, reports of the day's events are conveyed to the viewing public by way of alternate universes. The Fox News cable channel conveys its version of reality, while at the other end of the ideological spectrum MSNBC presents its version. They and their many counterparts on radio are more the result of an economic dynamic than a political one. Dispatching journalists into the field to gather information costs money; hiring a glib bloviator is relatively cheap, and inviting opinionated guests to vent on the air is entirely cost-free. It wouldn't work if it weren't popular, and audiences, it turns out, are endlessly absorbed by hearing amplified echoes of their own biases. It's divisive and damaging to the healthy functioning of our political system, but it's also indisputably inexpensive and, therefore, good business. And cable television and talk radio remain models of objectivity and restraint compared to what is routinely exchanged as "information" and "news" on the less restrained regions of the Internet. Even as digital tools elevate worthy voices once shut out of mainstream civic discourse, the Internet is also giving rise to "filter bubbles" that decrease users' exposure to conflicting viewpoints and reinforce their own ideological frames. As these various opinion echo chambers grow in influence, the broadcast news networks continue to lose audience and traditional newspapers struggle to survive.

Daniel Patrick Moynihan, the late senator from New York, once famously suggested that while everyone is entitled to his own opinion, he's not entitled to his own facts. The implication was that opinions could be, indeed should be, modified by facts. Nowadays, though, when facts themselves frequently become the object of partisan disagreement, opinions simply calcify into certainties. Attempting to alert the American public to an impending crisis becomes more difficult when the subject itself is complicated and defies easy or brief explanation. If only we could defer to the experts—but in today's political environment we have become conditioned to the notion that there is an expert to support almost any point of view. It has never been more difficult to convince the American public of anything that it is not already inclined to believe.

Ours has become a largely reactive culture. We are disinclined to anticipate disaster, let alone prepare for it. We wait for bad things to happen and then we assign blame. Despite mounting evidence of cyber crime and cyber sabotage, there appears to be widespread confidence that each can be contained before it inflicts unacceptable damage. The notion that some entity has either the ability or the motive to launch a sophisticated cyberattack against our nation's infrastructure, and in particular against our electric power grids, exists, if at all, on the outer fringes of public consciousness. It is true that unless and until it happens, there is no proof that it can; for now, what we are left with, for better or worse, is the testimony of experts. There will be more than a few who take issue with the conclusions of this reporter that the grid is at risk. This book reflects the assessment of those in the military and intelligence communities and the academic, industrial, and civic authorities who brought me to the conclusion that it is.

On April 13, 2010, a bipartisan group of ten former national security, intelligence, and energy officials, including former secretaries of defense James Schlesinger and William Perry, former directors of central intelligence John Deutsch and James Woolsey, and former White House national security advisors Stephen Hadley and Robert McFarlane, sent a confidential letter, not previously released, to congressman Ed Markey, the chairman and ranking member of the House Committee on Energy and Commerce. Written in support of the pending Grid Reliability and Infrastructure Defense Act, the letter came to some blunt conclusions: "Virtually all of our civilian critical infrastructure—including telecommunications, water, sanitation, transportation, and healthcare—depends on the electric grid. The grid is extremely vulnerable to disruption by a cyber- or other attack. Our adversaries already have the capability to carry out such an attack. The consequences of a large-scale attack on the U.S. grid would be catastrophic for our national security and economy."

It went on to say: "Under current conditions, timely reconstitution of the grid following a carefully targeted attack if particular equipment is destroyed would be impossible; and according to government experts, would result in widespread outages for at least months to two years or more, depending on the nature of the attack."

The House passed the proposed legislation. It has been stuck in the Senate ever since.

AK-47s and EMPs

This wasn't an incident where Billy-Bob and Joe decided, after a few brewskis, to come in and shoot up a substation.

—A FORMER VICE PRESIDENT FOR PG&E

While cyberattack is the most serious threat to our electric power system and is the primary focus of this book, it is not the only threat. Later in this chapter we'll examine the impact of an electromagnetic pulse, or EMP, attack. If it sits at the far end, or least likely end, of the spectrum, the simplest and, therefore, most likely form of attack has already taken place. Who was responsible and why remain, at this writing, as much of a mystery as when the attack took place. We cannot even conclude with any confidence that the attackers intended to inflict maximum damage. Whatever the goals of the attack, it provides an introduction to the shortcomings of grid security.

The scene was shortly before 1:00 a.m. on April 16, 2013, at the Pacific Gas and Electric Company's Metcalf Transmission

Substation, a few miles south of San Jose, California. To understand what happened, we rely on the exhaustive investigation by *Wall Street Journal* reporter Rebecca Smith. It is important to note at the outset that since her story was published in February 2014, no authority has questioned the accuracy of her work. We know that there were several saboteurs, but not how many. At least two members of the unit lifted a metal vault cover (too heavy for a single individual) leading to an underground vault containing AT&T's fiber-optic telecommunications cables. With the cutting of those cables, the attack began.

Slightly more than half an hour after cutting communications, the saboteurs attacked the actual substation, knocking out seventeen giant transformers over the course of nineteen minutes. Based on shell casings found at the scene, investigators believe that the gunmen used AK-47 assault rifles. In a remarkable feat of timing or coincidence, the saboteurs left the scene at 1:50 a.m., just one minute before the police arrived to find the substation locked. Video from surveillance cameras was of little help because the cameras were aimed toward the substation, while the shooters were positioned outside the perimeter.

The Metcalf substation provides power to Silicon Valley. Electric grid officials were able to avoid a blackout by rerouting power and calling on other plants in the region to provide additional power. The attack caused significant damage—it took utility workers twenty-seven days to bring the substation back online—but hardly the catastrophic result such a coordinated attack might have produced. These attackers seemed to know what they were doing. As a former vice president of transmission for PG&E told a utility security conference seven months after the attack, "This wasn't an incident where Billy-Bob and

Joe decided, after a few brewskis, to come in and shoot up a substation. This was an event that was well thought out, well planned and they targeted certain components." Still, if the attackers' goal was a widespread regional power outage, it failed. This interpretation aligns with industry claims that the power grid is far more resilient than critics suggest.

Jon Wellinghoff, who was chairman of the Federal Energy Regulatory Commission (FERC) at the time of the attack, remains unconvinced. He thinks the attackers may have been engaging in a rehearsal rather than a comprehensive sabotage operation. While he was still chairman of FERC, Wellinghoff assembled a team of experts from the U.S. Navy's Dahlgren Surface Warfare Center, which trains Navy SEALs, and took them out to the Metcalf substation. What they found, among other things, was that the shell casings left behind were free of fingerprints. They discovered small piles of rocks at key locations outside the substation and concluded that these might have been placed by advance scouts, establishing the most advantageous shooting locations. The experts concluded, as Wellinghoff told the *Wall Street Journal*, that "it was a targeting package just like they [SEALs] would put together for an attack." Wellinghoff's concern is that the attack on the Metcalf substation may have been a dry run for a far more devastating act of sabotage. Wellinghoff cited an analysis by FERC concluding that if nine of the country's most critical substations were knocked out at the same time, it could cause a blackout encompassing most of the United States.

In a conversation almost two years after the event, a senior executive for one of the nation's largest electric power companies dismissed Wellinghoff's comment as "idiotic." Was it "idiotic,"

I asked, because he shouldn't have made the statement, or because it wasn't true?

"Both," the executive replied.

We may never know whether the Metcalf attack was intended as a full-scale effort to disable part of the western grid or whether it was designed to pave the way for some future attack. In this instance, at least, damage to the grid was contained in relatively short order.

Understandably, the agencies most attuned to defending against attacks of all kind reside within the military. The North American Aerospace Defense Command (NORAD) is, in many respects, the nation's first line of defense. During the height of the Cold War, there was a greater public awareness of NORAD than there is today. In the event that Soviet bombers were headed our way, NORAD would provide the first alert, and U.S. fighter planes and bombers would scramble based on that intelligence. More ominously, NORAD would have also provided word if Soviet missiles had left their silos, giving the president and other top U.S. decision makers something less than half an hour before the missiles hit. During those thirty minutes they would have to decide whether to launch the first wave of retaliatory strikes. These days, NORAD's peak period of public visibility is at Christmastime when it tracks the course of Santa and his sleigh. Lower visibility notwithstanding, NORAD remains actively engaged in the nation's defense.

In early April 2015, the Pentagon, in a move that received hardly any public attention, announced a $700 million contract with the Raytheon Corporation to relocate critical computer systems deep underground into the massive bunker under Cheyenne Mountain in Colorado. These included the communications gear—computers, sensors, and servers—of the U.S. Northern Command (NORTHCOM), which is tasked with providing homeland defense, as well as the electronic gear serving NORAD, which continues to provide aerospace and maritime early warning against enemy attack. The Cheyenne Mountain bunker had been built back in the 1960s to withstand a Soviet nuclear missile or bombing strike, but now it was being modernized to withstand a different kind of attack. Admiral William Gortney, who in December 2014 took command of NORAD and NORTHCOM, explained that the move was designed to shield the electronic communications gear from an extreme solar storm or from an electromagnetic pulse (EMP) attack.

The potential impact of an EMP attack is so disastrous that it makes even the potential consequences of the Metcalf substation attack pale by comparison. It is literally the stuff of postapocalyptic fiction, receiving significant popular attention with the publication of William Forstchen's 2009 novel, *One Second After*. In Forstchen's telling, the Iranians and North Koreans have launched nuclear armed missiles from container ships off the coast, exploding them at high altitude over the United States. The resulting electromagnetic pulses have wiped out all electric power across most of the country, with truly horrific consequences for the entire nation. To all intents and purposes, and setting aside the issue of which nation might do it, that is how an EMP attack would likely be launched. It differs from any other nuclear attack in that its destructive power lies not in its radioactive fallout or in the physical destruction of population and structures but in its destruction of electronic equipment over an extremely wide area.

Forstchen's novel focuses on the consequences of such an

attack, exploring how the community of Black Mountain, North Carolina, might approach the struggle for survival amid growing evidence that other parts of the country have totally disintegrated. While his vision is well researched and convincingly imagined, it is, in the end, fiction. Yet it follows closely the findings of a congressional commission tasked in 2008 with identifying what the impact of an EMP attack would be on our civilian infrastructure. The commission's report, available online, outlines in matter-of-fact fashion the technical vulnerabilities of the power grid and predicts the likely impact of an EMP attack. Some projections are so extreme as to effectively numb the brain. There is simply no reasonable way to respond to those few lines in the commission report estimating that only one in ten of us would survive a year into a nationwide blackout, the rest perishing from starvation, disease, or societal breakdown.

When asked to identify all hostile actors who might launch an EMP attack by 2023, the commission concluded that there were several nation-states capable of such an attack; even more alarming was its conclusion that such an attack could be carried out by a terrorist organization. But why would any of the world's nuclear powers consider the detonation of an EMP device when the specter of nuclear attack has conjured up such alarming visions of retaliation and widespread destruction that the option has largely been shelved?

Former director of central intelligence James Woolsey, who warned of the growing threat of an EMP attack in a 2014 *Wall Street Journal* column, argued that capability is reason enough for concern, given how aggressively it's being pursued. "Rogue nations such as North Korea (and possibly Iran)," wrote Woolsey, "will soon match Russia and China and have the primary ingredients for an EMP attack: simple ballistic missiles such as Scuds that could be launched from a freighter near our shores." Particularly since Iran and North Korea still lack the capability to reach the continental United States with their intercontinental ballistic missiles, an EMP attack would give them a nuclear option. Woolsey filled in some of the blanks as to how the North Koreans may have acquired their EMP expertise, revealing that in 2004 Russian military personnel warned the EMP commission that North Korea had recruited Russian scientists to develop its nuclear and EMP attack capabilities. Woolsey contended that back in late 2012 the North Koreans successfully orbited a satellite capable of delivering a small nuclear warhead. Designated the KSM-3, this North Korean satellite could, said Woolsey, deliver a surprise nuclear EMP attack against the United States.

If Woolsey is correct and Russian scientists have transferred their knowledge of nuclear and EMP technology, the former CIA director's concern is understandable. North Korea, rogue state that it is, would stand out as one of the only regimes in the world whose rational restraint cannot be taken for granted.

James Woolsey argued that protection of the national electric grid against an EMP attack is possible, that it is not prohibitively expensive, and that necessary congressional action is long overdue. In its 2008 report the EMP commission recommended that measures taken by the Defense Department to protect crucial military installations, including the installation of surge arrestors and Faraday cages, could usefully be applied to civilian infrastructure also. The commission estimated that protecting the national electric grid against an EMP attack would cost about $2 billion. Such estimates are easily made in the abstract, but reality is another matter. No action was taken on the commission's recommendations for protecting the electric power grid.

Two pieces of legislation have been generated by later congressional committees. In an apparent bid for most tortured acronym of the year, the Secure High-Voltage Infrastructure for Electricity from Lethal Damage (SHIELD) Act was introduced in June 2013. The Critical Infrastructure Protection Act was introduced in October 2013. Neither piece of legislation, noted Woolsey, has made it out of committee. It is unclear whether our elected representatives have decided that the threat of an EMP attack is not that realistic after all or whether the failure to act owes more to their conclusion that there are more pressing issues requiring the expenditure of more than $2 billion. In the endless competition for federal funding, Washington has grown inured to the chorus of lobbyists crying wolf on behalf of one cause or another.

Those charged most directly with protecting the nation's security are ambivalent about the actual danger of an EMP attack. Janet Napolitano, the former secretary of homeland security, all but dismissed the threat. "You know, if I had to rack and stack the most likely risks that we were dealing with," she told me, "I would not put that [an EMP attack] in the top, certainly not in the top ten threats to our infrastructure."

Meanwhile, as noted, NORAD and the nation's Northern Command are moving their most sensitive communications equipment to a bunker below Cheyenne Mountain. Conflicting risk assessments among our national leaders and foremost experts will become a recurring theme.

Regulation Gridlock

It's Congress screwing around that you
should be mad at.

—SENATOR JOHN D. ROCKEFELLER IV

One hot summer afternoon in August 2003, a high-voltage power line in northern Ohio brushed against some overgrown trees and shut down. Sagging power lines, softened by a combination of weather and high current, are a common enough occurrence, and ordinarily the problem would have tripped an alarm. On this occasion the system failed.

As system operators struggled to diagnose the problem, three other lines failed in the same way, forcing the surrounding grid to take on additional current. In just an hour and a half these overburdened lines fell like dominoes, resulting in the largest blackout in North American history. Fifty million people lost power for up to two days in an area that spanned southeastern Canada and eight northeastern states. Eleven deaths were attributed to the outage, and repairs cost an estimated $6 billion.

A U.S.-Canadian task force, headed by senior advisors to the U.S. Department of Energy (DOE)—the chairman of the Federal Energy Regulatory Commission and the Department of Natural Resources, Canada—and chaired by the U.S. secretary of energy and the Canadian minister of natural resources, was established to investigate the blackout. It found that a number of parties had failed to adhere to industry standards. That bland, bureaucratic phrase, "failure to adhere to the industry's standards," drew critical attention to the embarrassing fact that in 2003 the electric power industry's standards were entirely voluntary. Companies that chose to ignore them were entirely free to do so. The U.S.-Canadian task force recommended the implementation of mandatory standards; we'll look at the mixed results of that recommendation a little further on. It also unexpectedly focused particular attention on a new, seemingly unrelated weakness. The grid, the task force found, was vulnerable to a malicious cyberattack.

It was a remarkably prescient conclusion, considering that the blackout of 2003 was in no way the result of a cyberattack. Following a three-year investigation of North America's most devastating power outage ever, the task force was warning that hackers could inflict even greater and longer-lasting damage. The industry took note and has since taken modest steps toward accepting mandatory cybersecurity standards. However, this is not an industry that easily embraces external mandates of any kind.

Imagine if an enemy of this country had crafted a system exposing the United States to the most devastating attack in its history. We may already have done a better job of inadvertently designing just such a system ourselves, sowing the seeds of our own downfall more effectively than any enemy of ours could have done. The federal agencies best equipped to monitor infrastructure for signs of cyberattack are precluded from doing so by laws that were designed to preserve privacy. Where there are breaches of infrastructure security, corporations are protected by law against any mandate to share that information with competitors or the federal government. The Obama White House has pushed the notion of information sharing by private industry as essential to deterring cyberattacks; business interests have countered by stressing liability concerns. There is, quite simply, an unavoidable tension between industry's insistence that it be allowed to operate within a free enterprise system and government's responsibility to develop high standards of safety and security for what may be the nation's single most critical piece of infrastructure. This tension has resulted, in the electric power industry, in a high-stakes duel between corporations and government regulators, the consequences of which are cybersecurity regulations so patchwork and inadequate as to be one of the chief sources of the grid's vulnerability.

The vast majority of electric power companies are privately owned. This has always been true, but a generation or so ago, that majority was made up of relatively few companies. In a model known as vertical integration, these companies owned the plants that generated the power, and they also owned the transmission facilities and the equipment that ultimately delivered the electricity to schools, businesses, hospitals, and homes. The managers of these industries had to be equally invested in securing the equipment that generated the power, safeguarding the transformers and lines that transported it, and protecting the hardware that delivered it to their consumers. Electricity was, then as now, generated by a variety of means, including nuclear, coal, natural gas, and hydro; whatever the energy source, the business model was essentially that of a monopoly, whose interests demanded attention to every aspect of the system.

Slowly, in one state after another, that system has given way to one characterized by limited competition. Now, power is often generated in one location by one company, fed over a separately managed transmission network (often overseen by either a regional transmission organization [RTO] or an independent system operator [ISO]), and ultimately passed on to yet another company for final delivery to the consumer. The consumer has no relationship with the companies that generate electricity or those that transmit it at top efficiency across vast distances. The consumer deals directly only with the local company that delivers the electricity on that final leg. The fact that there are now so many companies competing up and down the line is, of course, what produces a more aggressive marketplace, which works to the advantage of the consumer. That's the good news.

Because the system's maintenance and protection reside in so many different hands, though, and because its complexity has made each player more dependent on computerized control systems, the grid is also more vulnerable than it used to be. New forms of interconnections between and among firms create new pathways through which malicious cyberattacks may travel. Security and day-to-day reliability become a shared responsibility, and as with any other chain, the electric power grid may only be as strong as its weakest link. Leaders in the industry will argue that they have invested enormous resources in protecting their infrastructure, and they have. But smaller companies with lean profit margins are simply not inclined to spend a great deal on cybersecurity. The weakest links in this system tend to be the smaller companies with the poorest security and maintenance practices.

General Keith Alexander, who retired as director of the National Security Agency (NSA) in 2014, explained the issue as a simple cost/benefit ratio. "Your small and medium-sized companies cannot afford a world-class cyber threat team," he said. Major corporations might have "world-class" teams, but "you go to a small bank and they say, 'I can't afford that. I'm trying to make it through next week, and you want me to buy some cyber guy Google is trying to pay? I can't do that.'" This presents a particularly serious problem within the power industry because of the interconnectedness of the grid. "If you bring down the small [companies] in the right order," Alexander explained, it could initiate a domino-like "cascade effect." Cascading outages could compromise the systems of larger companies, quickly threatening the entire network. "It's not that they're bad," he said of less well-defended companies. "It's just that they don't have the infrastructure, the resources to do what actually needs to be done."

One might assume that the federal government, in the interest of safeguarding what is arguably the most critical infrastructure network in the country, can simply impose security and maintenance standards on the industry. But at present it cannot.

The grid has been operating according to reliability standards since the 1960s, but until 2003 those standards were established by the industry's own membership and compliance was entirely voluntary. There was the impression of oversight, but this was an illusion. Until 2003 the industry's coordinating body, the North American Electric Reliability Corporation (NERC), merely suggested standards to its membership. No sanctions were imposed on companies that ignored them.

The August 2003 blackout was a wake-up call for both industry and government regulators. What ultimately emerged, though, still amounts to a lot less than effective oversight of the electric power industry. Regulations that were once optional are now mandatory. That's a significant change, but the industry continues to have the last word on which of the regulations put forward for the governance of its conduct it is prepared to accept. The Federal Energy Regulatory Commission can and does propose regulations for the industry. But each of those proposals has to be put to a vote by the NERC membership, which represents the entire industry. Since each regulation proposed by FERC requires approval by two-thirds of NERC's members— large and small, major corporations and those with tighter budgets—the system is not designed to generate the most rigorous standards.

Only when the industry has shaped and polished the regulations that will govern its behavior is FERC finally empowered to enforce the rules that emerge. NERC has an enforcement capacity as well, but in 2013 it levied only $5 million in penalties against utilities failing to meet mandatory defense standards. In 2014 the sum of all penalties dropped to less than $4 million. To provide context, in August 2013 the industry's 3,200 utilities sold $400 billion worth of electricity; penalties added up to less than 0.001 percent of gross revenues. Those wishing to view the glass as half full will point out that electric power is one of the few infrastructure sectors where mandatory cybersecurity regulations of any kind have been established. However, when all is said and done, FERC is merely enforcing regulations that it may have proposed but which the industry itself has modified to meet its own interests.

Then, too, there is FERC's limited jurisdiction. American democracy rests on a foundation of competing tensions among local, state, and federal laws, and laws governing the electric power industry reflect those tensions. The nationwide transmission of electricity along high-voltage power lines is subject to federal regulations, enforced by FERC. Once the electricity has been conveyed from the generating facility to its ultimate point of distribution, though, it is no longer under federal jurisdiction. To be clear: while electricity is being conveyed across the country at maximum efficiency, the process is subject to federal regulations. Once it's handed off to the local companies that transmit electricity to the local consumers, no federal regulations apply. At that point, it's under state authority—fifty different jurisdictions, in which regulations focus almost exclusively on the financial side of the business. The state commissions that govern those local companies control the rates that local utilities can charge, but many of them pay scant attention to issues of grid security. Yet security standards are legislated as a state-by-state responsibility, and the industry has taken the position that any state law insisting that the industry itself bear that liability, from the point of generating electricity up to and including the point of delivery to the consumer, would conflict with federal law. Once the electricity gets to its actual point of distribution, the industry argues, any federal laws governing the industry no longer apply.

To say that this loophole sticks in George R. Cotter's craw is to understate his passion. His expertise on the vulnerabilities of the electric power industry comes from a lifetime's experience within the National Security Agency, in which he served as chief of staff, as chief scientist, and twice as head of its technology division.

Cotter fumes at what he regards as the idiocy of the regulations enforcing the security of the electric power industry. The final leg of the system, Cotter explained, everywhere that electricity is actually delivered to consumers, is not covered by federal security regulations, and the industry wants to keep it that way. "Most of the critical infrastructure is in urban areas. Entire national security establishments are not covered by the law or by NERC." Of his old NSA headquarters, he noted that while the agency itself is protected, "the electric power coming into the agency is not." When asked about the chances of industry supporting a national set of standards, Cotter was pessimistic. "They view deregulation as far more important than cybersecurity protection for their facilities," he said, echoing Keith Alexander's business-first explanation. "They've taken a very strong position on this with FERC, that the communications and networks that interconnect all of this are not covered by the standards."

Congressman James Langevin, a Democrat from Rhode Island, is cofounder and cochair of the Congressional Cybersecurity Caucus. He sits on both the House Armed Services Committee and the House Intelligence Committee. What keeps him up at night is the fear of an attack on critical infrastructure and the inadequacy of current security measures. He knows how firmly the security of the electric industrial grid is in the hands of the private sector. "Nobody's in charge there, nobody has responsibility, nor can anybody require that they do work. One would think that FERC could direct and require more cybersecurity be employed by the owners and operators of the electric grid. They do not."

Former senator John D. Rockefeller IV (D-W.Va.), who has served as chairman of both the Senate Commerce and Intelligence Committees, has similarly given up hope of enforcing tough security measures on the electric power industry. There was a time, he told me, when he could count on the support of a number of senior Senate Republicans. "Then all of a sudden comes the Chamber of Commerce in 2011 and some lobbyist goes back there and says, 'We gotta shut this thing down . . . overregulation, heavy-handed government, et cetera.' "

"What were they afraid of?" I asked.

"They were afraid," said Rockefeller, "of having to spend money that they couldn't prove to themselves they would actually need to spend."

Rockefeller was wistful about the level of bipartisanship that existed before the Chamber of Commerce intervened. He ticked off the Republicans: Kay Bailey Hutchison, John McCain, Olympia Snowe, Susan Collins, Joe Lieberman. They would meet in a secure room at the U.S. Capitol Visitor Center. "We had so many meetings with so much braid. There would be this mass of generals and admirals, all saying, as Mike McConnell [then director of national intelligence] said: 'For the record, if we were attacked, we would lose.' "

"He was talking about cyberattack?" I asked.

"Yes," said Rockefeller.

4

....

Attack Surfaces

Here's that password.
—HACKER AT BLACK HAT CONFERENCE

It is not easy to convey how and why the electric power grid is so surpassingly vulnerable to cyberattack. First we must understand a little more about the exchange and conveyance of electricity. As discussed, deregulation of the electric power industry broke up the old, vertically integrated monopolies. Whereas those companies generated the electricity, sent it across great distances along high-voltage transmission lines, and then distributed that electricity to the consumers, now different companies are responsible for different phases of the process. The company that distributes electricity in one community, for example, can buy power from a number of companies generating electricity in other parts of the country. Breaking up the industry into a marketplace of interconnected parts introduced competition, which lowered prices. It also increased the system's vulnerability to cyber intrusion.

With apologies to what is an infinitely more complex and sophisticated industry, imagine a giant balloon attached to a thousand different valves. Some of the valves introduce air into the balloon, while others extract it. Take too much air out and the balloon collapses. Put too much in and it explodes. Now imagine a computerized system that keeps it all in balance, so that the balloon remains perfectly inflated. That's a very crude analogue to the system that keeps the electric power industry in balance. If you could hack into that computerized system and throw supply and demand out of balance, it could have devastating consequences.

These days, Richard A. Clarke is chairman of Good Harbor Security Risk Management, a Washington, D.C., firm that offers advice on cybersecurity. In 1998 President Bill Clinton named Clarke national coordinator for security infrastructure protection and counterterrorism. He remained at the White House in the same role under President George W. Bush, eventually becoming special advisor to the president on cybersecurity. I turned to Clarke because he is particularly adept at reducing complex technical issues into layman's language. "So it's winter," said Clarke, "and Chicago needs more electricity" between the hours of, say, 2:00 and 5:00 p.m. "Florida Power and Light, which has more capacity than it needs in the winter, says, 'I can provide it and here's how much I'll charge.'" Assume that Florida Power and Light has offered the best deal, to the advantage of consumers in Chicago. But now that electricity has to get from Florida to Chicago, and pathways are limited. Communities are inclined to object to the presence of high-tension wires near people's homes. Because of what's come to be known as the NIMBY (not in my backyard) factor, there are relatively few high-power transmission lines in key locations around the country.

Clarke likes to compare those high-power transmission towers and cables to a rail line. It's an analogy that shouldn't be taken too literally, other experts have cautioned, but it is helpful in underscoring the fact that the conveyance of electricity along those transmission lines has to be scheduled. In our theoretical deal, Florida Power and Light has agreed to dump power onto the grid between two and five in the afternoon. The difference between trains and electricity is that electricity isn't conveyed directly from point A to point B. The electricity leaving Florida probably won't be delivered to Chicago. It's all about maximum efficiency and maintaining overall balance between demand and supply throughout the system. To coordinate this, the industry has set up regional authorities, the regional transmission organizations and independent system operators, which monitor "traffic" to ensure that no transmission lines in their area become overburdened.

This monitoring process, while routine, also creates a dangerous point of vulnerability. If someone was able to hack into an RTO or ISO and deliberately overload the lines, the impact would be swift and physical. The lines would start to droop from the heavy load. They would overheat. "When the lines dip," said Clarke, "they can set a tree on fire, or they can melt the line." There are built-in controls to ensure that such an overcapacity never happens, but if a hacker got into the system and targeted those controls, Clarke explained, so that "the guy sitting in the operations center doesn't see it—he sees that everything is in the green," there would be no relationship between the operations center dashboard and reality. Such a situation could quickly escalate out of control. If you can break key transmission lines, said Clarke, you can produce cascading, potentially catastrophic outages.

Is it doable? It is anything but simple. It would require detailed mapping and lengthy reconnaissance operations to conclude what to target and how to find a critical point of failure in the system. But it is, technically speaking, more plausible today than ever before. Deregulation of the power industry has created a system with more vulnerable points of entry than ever existed previously, and a lot of the equipment is controlled by aging, standardized computer systems used around the world and familiar to many of America's enemies.

Many of the old power stations were operated by manual controls. If they had any computer software at all, it was unique to that company. These days, within any one of the three U.S. grids, almost all operational phases of thousands of power companies are interconnected. Coordinating operations are run using the same supervisory control and data acquisition (SCADA) systems. Most of the systems are manufactured by a relative handful of companies, and while they are not quite interchangeable, there are similarities in programming and structure. This presents a web of pathways connecting the thousands of power companies and enabling transactions such as that Florida-to-Chicago transfer. The overall system has been designed for maximum efficiency, eliminating waste while establishing a precise balance between the power needed and the power generated.

Craig Fugate, administrator of the Federal Emergency Management Agency, is concerned that we have sacrificed resiliency in the interest of achieving efficiency. "We have created a system," Fugate told me, "where we generate power in very efficient quantities at specific locations, and then we have to move that power, oftentimes at great distances, to where it's being consumed." If someone was knowledgeable about the functioning of a SCADA system and succeeded in hacking into it, that individual could engineer "a series of events that seem totally unrelated" but which could, according to Fugate, "turn the lights out very quickly over large areas."

Richard Clarke agrees. "If you go into a big, modern power station in Shanghai, or a big, modern power station in California, you're going to find the same SCADA software." SCADA systems were, for the most part, designed and installed before the notion of cyberattacks had even occurred to anyone. The Internet itself was not designed to keep anybody out. It was created to be universally accessible.

There's a small historical irony in the fact that the extreme vulnerability of SCADA systems to cyber sabotage was first demonstrated, and dramatically so, during the administration of George W. Bush. At about the same time that Richard Clarke was serving as White House advisor on cybersecurity, U.S. computer specialists at the National Security Agency and their counterparts at the Israeli military's Unit 8200 launched a cyberattack against a critical element of Iran's nuclear program. Iran's nuclear program was set back by as much as two years, according to some estimates.

The attack, code-named Olympic Games, targeted an array of several thousand nuclear centrifuges located at Natanz, Iran's main enrichment center. These centrifuges spin uranium gas at the high speeds necessary to refine the uranium used to fuel both nuclear reactors and bombs. With the introduction of a computer worm code-named Stuxnet, the cyber saboteurs were able to alter the speed of those centrifuges, undermining the refinement process.

Sending the centrifuges into a destructive spiral would have been only marginally damaging had the Iranians recognized what was happening. They could have responded in time to mitigate the attack. The genius of the U.S.-Israeli attack lay in its ability to conceal the sabotage. According to David Sanger, who reported on Olympic Games for the *New York Times,* Stuxnet "also secretly recorded what normal operations at the nuclear plant looked like, then played those recordings back to plant operators, like a pre-recorded security tape in a bank heist, so that it would appear that everything was operating normally while the centrifuges were actually tearing themselves apart."

The SCADA system controlling those nuclear centrifuges in Iran was manufactured by Siemens, as is much of the SCADA software used by the electric power industry in the United States. All of that Siemens software has a built-in access point accessible only to Siemens engineers employing a closely held password. Or so Siemens believed. As Richard Clarke recounted to me, one attendee got up at the 2011 Black Hat hackers conference in Las Vegas and announced: "Here's that password." Siemens had to go to every deployment of their software worldwide and change the password.

Getting into a piece of critical infrastructure is one thing, but it's worth repeating that navigating an electric grid is a highly complex operation. The reconnaissance required to understand the system sufficiently to compromise it can take years, challenging the skills of even the most cyber-competent nation-states. We'll get into what the experts call "preparing the battlefield" in a later chapter. (Several nation-states, most prominently the Russians and the Chinese, have already spent years conducting just such reconnaissance.) For the moment, suffice it to say that it's difficult to keep hackers out of the system.

Analogies are imperfect, but they can be instructive. During the fall of 2014 we were reminded incessantly that terms we have grown accustomed to associating with the Internet were, in fact, borrowed from the field of medicine. The shared terminology is not accidental. How better to describe the spread of an alien intruder through a computer program than as a "virus"? "Hygiene," "anti-virus protection," "immunity," and "vaccination" are terms commonly used in both medicine and cyberspace; each refers to a defense against "bugs." Searching for the origin of a virus, for "patient zero," is strikingly similar to examining the outer edges of a network that has been infected by malware. One failure to disinfect, to follow the proper protocol, and a virus spreads from carrier to carrier or program to program.

When Thomas Eric Duncan, infected with the Ebola virus, was finally admitted to Dallas's Texas Health Presbyterian Hospital in the fall of 2014, he was placed in isolation, where he was treated by nurses Nina Pham and Amber Vinson. Each nurse wore two gowns, two pairs of gloves, shoe covers, a surgical mask, and a face shield. It was not enough. Duncan was in the worst throes of the disease, subject to explosive diarrhea and projectile vomiting. Either because some of Duncan's bodily fluid splashed onto an exposed portion of their neck or because they inadvertently touched some of the fluid while taking off their gloves, Pham and Vinson became infected with the Ebola virus. Both women ultimately recovered, but in each case, the tiniest margin of error provided an opportunity for infection. In the context of cyberattacks, that vulnerable spot, the equivalent of the small exposed portion of the neck, is what specialists call an "attack surface"—a vulnerable entry point to what is believed to be a secure operating system.

In terms of the power grid, the number of attack surfaces has increased exponentially with the integration of everyday devices on the Internet. Whereas a prospective hacker in the past might have had to go after a server or a desktop computer to gain access to an electric company's corporate network, now he can do it by way of the devices that enable a consumer to program the lighting or heating and air-conditioning in his home remotely or automatically. The "smart" thermostat that automatically lowers the temperature in a customer's home at night or warms his kitchen before he gets up in the morning has to be connected to the company's billing department, which in turn needs to be connected to whatever department actually conveys electricity to the home. Each connection provides another potential attack surface.

In theory, the administrative network is "air-gapped" from the operational side of each power company, meaning that there is no physical connection between the two. Power companies insist that those two networks are absolutely separate and not connected. Whenever Homeland Security or the Federal Energy Regulatory Commission has hired computer forensic experts to investigate this claim, however, they have found minute connections. A Verizon/Secret Service study concluded that two-thirds of companies across a spectrum of industries didn't realize they had been breached until someone outside the company informed them. Another study, conducted by the cybersecurity firm FireEye, found that it took on average 279 days before companies that had been breached came to realize it or were told by someone else.

The problem with air-gapping, one academic specialist warned me, is that it fails to take the human factor into account: "Every time a worker brings in a thumb drive or laptop from home and hooks it up to an 'isolated' system, the mobility of workers bridges the air gap." As workers and users of the two systems transfer work to their personal computers at home, or from their smartphones or other interconnected networks back to what is supposed to be an isolated, secure system, they run the risk of infecting the operational network. Would-be hackers, operating on what is sometimes referred to as the "Sneakernet," can introduce their malware, their viral programs, by way of an employee's insecure iPhone or thumb drive. The insecure thumb drive becomes the cyber equivalent of the tiny exposed portion of the nurse's neck. Anything less than absolute hygiene provides a potential attack surface.

Absolute hygiene may be theoretically possible, but it would be prohibitively expensive—not just for smaller, local companies but also for regional authorities tasked with monitoring vast areas. PJM, for example, is a regional transmission organization that handles the output of more than thirteen hundred generating sources, which it then distributes to more than six thousand transmission substations, serving fourteen states on the Eastern Interconnection grid. George Cotter, former chief scientist for the NSA, explained that the larger RTOs use literally tens of thousands of devices, many of which are insecure, and all of which are interconnected. "A company like PJM," said Cotter, "isn't going to buy seventy thousand crypto devices. They'll wait until the companies that manufacture them build cryptography into them, and that's decades."

Time is not on our side.

Guardians of the Grid

We can't defend against everything, but right now
we're vulnerable to just about everything.
—MAJOR GENERAL BRETT WILLIAMS, FORMER DIRECTOR OF
OPERATIONS, U.S. CYBER COMMAND

Certainly no one has a greater interest in protecting the security
of the electric power industry than the industry itself—if only
cost were not a factor and profit were not an essential ingredient of staying in business. It is not altogether reassuring, then,
to consider that the only institution with real power to decide
how the power industry is protected is the power industry. In
evaluating industry regulations in 2012, the nonpartisan Congressional Research Service questioned the entire arrangement,
calling it "unusual" and observing that it "may potentially be
a conflict of interest" for an industry to legislate its own standards. Corporate leaders dismiss this, arguing that government
lacks the expertise to run the industry. On the other hand, they
are open to working with government agencies in an advisory

or supporting role, given the government's undeniable expertise in computing technology and security. It is, say industry representatives, a perfect partnering opportunity.

Three times a year, the Electricity Sector Coordinating Council, led by chief executive officers of the power industry, meets with senior administration officials. The council arranges for the deployment of security tools developed by the Defense Department, the National Security Agency, the Pacific Northwest National Laboratory, and the Department of Homeland Security's Science and Technology Directorate. While these agencies do pass along new technology when available, in large part what they convey to the power industry is information—reports on risk assessment designed to improve the industry's situational awareness and to help it detect as quickly as possible the presence of anomalies in the system.

It sounds encouraging: making sure that the top people in industry are getting the most up-to-date information at the right time. But as Major General Brett Williams explained, progress is halting. As the director of operations for U.S. Cyber Command until his retirement, Williams speaks with considerable authority. He warns that while the United States has the best cyber offense in the world, the same is not true of the nation's cyber defense. When he left Cyber Command in the spring of 2014, threat information was still being communicated between U.S. intelligence agencies and the electric power industry the "old-fashioned" way, via phone calls and emails. On the cyber battlefield, Williams explained, information needs to be communicated instantaneously. It's not that the technological ability to deliver automated, machine-to-machine warnings doesn't exist. Indeed, in 2013 President Obama issued an executive order making information sharing between critical industry and government cybersecurity services a priority. The order did not spell out the means by which information is to be communicated, however, nor is making something a "priority" a mandate. Not surprisingly, in fact, industry and government have different definitions of what constitutes a "priority."

On the intelligence side of the equation, it's sources and methods. Williams explained it this way: "If you look back at classic signals intelligence, you've got a bug in some foreign leader's office and you've got a piece of information there. If you use that information to make a decision, they're going to know that the only place that information could have come from is a bug in the office. We apply that same logic to things in cyberspace." We simply don't have the luxury, Williams argued, of continuing to place the protection of sources and methods ahead of instant response. Cyber defense demands speed. "I would argue we just don't have time to go through the same process that we go through with legacy types of intelligence. Obviously, we can't defend against everything; but right now we're vulnerable to almost everything."

Industry and government leaders alike recognize the threat, according to Williams. "There's plenty of people that understand what needs to be done. It's policy and it's money" standing in the way. The policy discussion currently under way, Williams explained, is establishing how far the Defense Department's role extends in defending the homeland against cyberattack. As things now stand, he told me, Cyber Command doesn't defend private industry; "the policy hasn't even matured enough to defend critical infrastructure." For any sort of cyber defense system to efficiently protect the electric power industry, information sharing has to be a two-way street. Corporations will have to get over their privacy and liability concerns and give government agencies the security data those agencies say they need in order to be effective. The military and intelligence agencies, in turn, need to make information relating to cyber threats available in real time, setting aside worries about jeopardizing sources and methods.

Scott Aaronson, national security director for Edison Electric Institute, the industry's trade organization, made it appear that the policy of sharing security data in real time was up and running. The program, as he later conceded, is actually in its infancy. Aaronson cited an initiative called CRISP, the Cyber Risk Information Sharing Program. All network traffic coming into the program from the outside, Aaronson explained, is fed through an ISD, or information sharing device. It doesn't block information; it simply analyzes the traffic, comparing it to classified and unclassified government databases. It can identify the IP addresses that need to be blocked and the malware that threatens secure operations. What's more, said Aaronson, it does all of this in real time.

A senior power company executive described what sounded like the same program in different terms. Cyber traffic designated as coming from "friendly" servers is "whiteboarded." Servers found to be connected with unfriendly foreign states, criminals, or hackers are tagged and blocked, or "blackboarded."

If that sounds a little too good to be entirely true, it is, just a little. If and when CRISP is deployed across the entire electric power industry, it will be an enormous step forward. But, as Aaronson acknowledged, creating a private-sector model around a government technology is not a simple task. As of early 2015,

CRISP was operating at what Aaronson called "near-real time." Machine-to-machine information was actually being shared between the government and fifteen companies—fifteen out of roughly three thousand, or, as he estimated, 0.2 percent of the entire industry.

How long, I asked, before CRISP is deployed across the entire industry?

"We will double this year. We will get to at least thirty [companies] by the end of 2015. I think there's a goal to be at forty." The information gathered by that tiny fragment is being "socialized" with thousands of other companies, but this is not happening in anything approaching real time and has not yet reached the point where it can be described as offering a serious defense of the industry at large.

It is difficult to focus the attention of power industry executives on speculative threats when there are so many existing problems to deal with. It can be almost impossible to convince investors or boards of directors to siphon off limited profits to prevent a crisis that, in their opinion, may never happen. However, industry leaders are at least pragmatic enough to consider the possibility of a successful cyberattack against the grid. Scott Aaronson acknowledged that "the mandate for perfect security implemented by imperfect network operators is a recipe for disaster," but he was not prepared to concede that a cyberattack could take down an entire electric grid. Industry representatives won't predict it. They don't expect it, they insist. But "I have been conditioned," said Aaronson, "to say nothing is impossible."

Doubts over security of the power grid cover a broad range, from that cautious "nothing is impossible" to the full-throated warning of "not if, but when." Uncertainty has given rise to a vast cybersecurity industry, with more than its share of distinguished former government employees. It is a long-established custom in Washington for men and women who have devoted some of their best years to the military, service in one or more administrations, congress, or an intelligence agency, to capitalize on their experience once they leave public service; these may have been whom Aaronson had in mind when he followed up his "nothing is impossible" comment with this additional observation: "But I am suggesting that [taking down a grid] is not nearly as simple as I think some people, who maybe have services they'd like to sell, would have people believe."

Among the experts I've consulted who now provide advice on cybersecurity in the commercial marketplace are two former secretaries of homeland security, two former White House advisors on cybersecurity, one former director of operations at Cyber Command, and one former director of the National Security Agency. None of these people has suggested that there is anything simple about sabotaging a grid; they say only that the capability exists, an opinion shared by specialists who have no financial interest in the field. When I asked Janet Napolitano, who is now president of the University of California, what she thinks the chances are that a nation-state or independent actor could knock out one of our power grids, she replied, "Very high—80 percent, 90 percent. You know, very, very high."

Still, the implication that the threat of a cyberattack against the grid has been exaggerated by people with "services they'd like to sell" was enough to make Richard Clarke bristle: "I don't sell to the electric power companies." As for the companies that do engage his services, "[they] call me after they've been breached. We get called up by people who have already had a problem, and that's every industry. It's big companies. It's banks that have spent hundreds of millions of dollars a year on it; still can't get it done. The idea that people like Keith [Alexander] and I are conjuring up cyberattacks which don't exist is laughable. I mean, every company that you can talk to has been breached. Just ask them."

I have tried, wherever possible, to keep all sources in this book on the record. It was difficult, however, to convince a top executive from one of the larger electric companies to publicly discuss the likelihood of a successful cyberattack on a power grid under any circumstances. When one finally did, I agreed to his condition that he be identified only as a senior executive for one of the nation's major power companies; let me add only that he speaks for the industry with considerable authority. He, too, was willing to concede that hackers can get into the system. "But," he added, "I almost guarantee you they're not going to be able to create widespread damage. In order to create widespread damage—I mean, when you say 'take down the grid,' you know, Long Island could go out. It could go black. But the rest of the United States wouldn't."

As noted in an earlier chapter, local distributors of electricity, even those in major urban areas, are not governed by any national regulatory standards, leading to the bizarre consequence that distribution of electricity—even to national security establishments such as the NSA, for example—is regulated only on a state-by-state basis. With no federal standards and fewer resources than the investor-owned corporations that generate and transmit power (the so-called bulk electric systems), those local companies tend to be more vulnerable. If enemies wanted to launch a truly devastating cyberattack, couldn't they go after the local distribution system?

It's a question I posed to Jim Fama, vice president of energy delivery at Edison Electric. Fama acknowledged that the distribution end has become more vulnerable as the entire system—from generating to distributing electricity—is increasingly digitized, but he argued that it was an unlikely target for a massive attack. "If you wanted to attack control mechanisms, utility control mechanisms," he said, "you would get a much bigger bang for the buck if you were to go after the bulk power system."

But that's precisely the point that cybersecurity experts such as Richard Clarke are making. Every time that Homeland Security or the Federal Energy Regulatory Commission has hired a forensic expert, he told me, they have found connections linking publicly accessible webpages to a given power company's administrative network, and through this to its operational network. "A lot of our companies," Fama insisted, "have isolated their SCADA systems and their EMS [electric management systems]. They're not connected to the billing and the customer information data. They're segregated." Note that Jim Fama's claim that operational networks are fully isolated is not for the entire industry, or even for his entire trade association, but only for "a lot of our companies."

The senior power company executive I spoke with heads one of the largest and, he insists, best-protected electric power companies in the country, if not the world. In our initial conversation he steered me away from the vulnerability of those SCADA systems. Where security matters most in the generation and transmission of power, he said—that is, in the bulk electric system—companies have upgraded their EMS. I mentioned that a number of top intelligence specialists believe that the Russians, the Chinese, and probably the Iranians are already inside the grid (as we'll examine in a later chapter). The executive rejected this notion out of hand. He did not believe, he told me, that anyone had cracked the EMS.

"If they have," he argued, "they would have done something. Especially Russia."

"Why?"

"Because they would want to disrupt commerce. They would want to disrupt American life. Do I think they're inside EMS? No, I don't. We go through extraordinary security there. We have air[-gap] protection. We regularly wipe servers."

Did I remember those Russian *matryoshka* or nesting dolls? he wondered. He was analogizing to a system in which no two operational parts are physically connected—while they sit one inside the other, they never touch. "One of the most valuable defenses is air. And that means that in order to go from A to B, there is no network. There is no physical connection to get through, and you have absolute security."

That, I pointed out, is just what the Iranian nuclear technicians believed as they watched television monitors showing normal operations, while in reality thousands of their centrifuges were spinning out of control. Air-gapping works unless and until an employee infects the system by bringing in a personal device—a thumb drive, say—from the outside.

"Oh, right. That's scary stuff," the executive conceded. He assured me, though, that there are "a lot" of standard tests, "both manual and automatic," to ensure that actual operations mirror system readouts. "The nature of the grid is such that, number one, its interconnectedness gives it more resiliency, and number two, we have automatic controls and protection regimes so that if problems develop we can actually isolate the problem."

Here was a top executive from one of the nation's largest power companies arguing that the grid's interconnectedness gives it more resiliency, even while government officials such as Craig Fugate, the FEMA administrator, had taken the diametrically opposite position. Fugate, remember, said that the electric power industry has sacrificed resiliency in the name of efficiency.

My conversation with the senior power company executive ended on a perfectly amicable note, but with minds essentially unchanged. "I don't mean to convey 100 percent confidence," he told me. "I'm just giving you what I believe." It was a bit of a surprise, then, when a few weeks after our conversation I received a call from that executive. Following our talk, he had chaired a meeting of the Electric Sector Coordinating Council. That's the group of power company CEOs who meet periodically with senior White House officials. The administration, he said, was particularly concerned about the vulnerability of the SCADA systems. "And I said, 'Yeah, I get SCADA'—just the same thing I said to you. 'I'd be much more concerned about EMS than I would SCADA,' and we got into a discussion on that." EMS, after all, is the system that directly controls the generation and transmission of electricity. "The meeting breaks up, and one guy, who I respect, came over to me and said, 'You know, SCADA could be a problem. We shouldn't dismiss SCADA.'"

Then the administration official asked the executive, "You remember Aurora?" He was referring to an incident in 2007 when the Idaho National Laboratory conducted something called the Aurora Generator Test. Staffers acting as hackers caused the circuit breakers on a giant diesel generator to open and close rapidly out of phase, in a process called "pinging," until the generator actually tore itself apart.

"He was afraid that if the bad guys got in [to a SCADA system] and did some sort of mass pinging attack, that you could create some problems, even before you knew you had a problem, and that that could provide imbalances in the big [EMS] system." It was an important proof of how much physical damage can be done through network hacking, and it raised the specter of how much greater the destruction could be if the network was penetrated more widely. For a cyberattack to inflict lasting and widespread damage, "it would have to be something massive," the senior power company executive maintained, "hitting lots of critical infrastructure somehow." But a high-ranking official—"somebody that I respect"—had just impressed upon him the reality of SCADA's vulnerability, so much so that he felt he "owed it" to me to acknowledge the seriousness of the threat. It was essentially the same thing that experts such as Richard Clarke had told me, but this was coming from the CEO of one of the nation's largest electric power companies.

It was an honorable admission to make. It also highlighted the fact that there is anything but unanimity within the industry on the issue of risk assessment.

What Are the Odds?

The industry is equipped to lose as much as two
hundred billion dollars. It would not be a great day,
but life would go on.

—AJIT JAIN, BERKSHIRE HATHAWAY

More than fifty years as a reporter, all too often as a war correspondent, have ingrained in me a healthy respect for the law
of unintended consequences: When the United States first sent
troops into Viet Nam in the early 1960s, or Afghanistan following 9/11, or Iraq in 2003, there was little expectation that those
"limited incursions" would expand into such life- and treasure-
sapping wars. Each conflict conjures its own images of unfore-
seen devastation. There is now ample evidence that the law of
unintended consequences applies as remorselessly in the realm
of cyberspace as it does anywhere else.

We live in a world of fine print and mellow-voiced warnings
gently bathed in soft music and issued over soothing, totally unrelated visual images. When the Internet first nuzzled its way

into our lives, we came to know it as "the Web"—an evocative concept, carrying the promise of a free exchange of ideas without the encumbrances of time or space. But a web can also trap, limit, and smother. It can be liberating and dangerous at one and the same time. And, of course, it has evolved.

The Internet was never designed to help criminals steal credit card data from the files of 110 million Target customers. But in 2014 it was used to do just that.

It wasn't designed to unleash a computer virus on Aramco's corporate PCs. But that's how it was employed back in August 2012, erasing spreadsheets and emails throughout the enormous Saudi company and replacing them, as then Defense Secretary Leon Panetta told an audience of security executives, with "an image of a burning American flag." The attack, Panetta reported, involved a complex virus called Shamoon, which rendered more than thirty thousand computers wholly useless. "In effect," one U.S. Army cyber specialist told me, "it turned those computers into thirty thousand bricks." It wasn't entirely clear at the time, but the National Security Agency has since concluded that the attack was Iran's answer to Stuxnet.

The architects of the Internet are not likely to have envisioned cyber criminals gaining surreptitious entry to the private files of a major bank, but JPMorgan's chief executive, Jamie Dimon, told shareholders in his annual letter of 2014 that "cybersecurity attacks are becoming increasingly complex and more dangerous." The bank, he noted, would be spending $250 million on cybersecurity that year.

In the spring of 2013 Dimon told me that JPMorgan had already spent more than $600 million on cybersecurity. Notwithstanding that vast and ongoing expenditure, Dimon's warning proved all too prescient: in the summer of 2014 hackers compromised eighty-four million JPMorgan files.

These and many other businesses have concluded that the advantages of the Internet are nevertheless worth whatever vulnerabilities may emerge as by-products. The electric power industry has made the same calculation. However dangerous the consequences of conducting our businesses and operating our infrastructure on the Internet, we are simply incapable of functioning without it. Quoted in a comprehensive *Washington Post* article titled "A Flaw in the Design," computer-science pioneer Peter G. Neumann neatly summarizes the problem of security on the Internet. "People always say we can add it on later. [But] you can't add security to something that wasn't designed to be secure."

There is precedent. It is hardly the first time that society has embraced a new technology without understanding its shortcomings. More than a century ago, who could have imagined an America teeming with more than 250 million cars and trucks, as it is today? How could our forebearers have anticipated traffic jams and air pollution? More to the point, perhaps, would our national leaders, state governors, city mayors, and town councils have permitted the development of the car if they had known that by the early 1970s more than fifty thousand of us would die on our roads in a single year? By the time that statistic became a reality, the bargain had been struck, sealed, and so interwoven into our culture, our daily lives, that no one would seriously propose eliminating the automobile. What we have since done is to reduce as best we can its potential for harm. It took a very long time before the industry acceded to the need for reduced speed limits, seat belts, air bags, and crumple zones. It took a coalition of enraged mothers whose children had died at the hands of drunk drivers to overcome the efforts of the alcohol industry and increase the penalties for driving while under the influence. We now endure a more modest twenty-five thousand highway fatalities annually. Still, we, as a nation, are prepared to pay a terrible, ongoing price in destroyed and damaged lives so that the rest of us may enjoy the advantages of progress.

There are isolated reports, in the wake of Edward Snowden's revelations about the global intrusiveness of the National Security Agency, that some parliamentarians in Germany and a branch of Russian intelligence are considering a return to typewriters and paper files. I read such stories wistfully but without any expectation that the movement will spread. The world is locked into a state of cyber dependency.

According to one military cyber specialist, "We think we're in total control of the computer, all the time. It does what I tell it and it's controlling all my pay, billing accounts; it's controlling hot water flow or the fuel flow. We have no visualization of what harm could be done by someone who has intruded into that system." He also pointed to a generational "security consciousness gap," noting that our children take the safety of the digital landscape even more for granted. "They expect immediate access to information, and they don't fathom that there's anything risky about it. So as the adversary's threat is going up, our security consciousness is going down."

The Internet, as we now know it, carries a trove of inherent dangers. Those dangers are just beginning to reveal themselves, and their scale and scope may someday call into question our easy acceptance of its benefits.

More than a hundred years ago, long before the power of the

Internet gave it the force of commonplace reality, Mark Twain commented on the uneven nature of any competition between rumor and fact, gossip and reality, observing that "a lie can travel halfway around the world while the truth is putting on its shoes." Twain was speaking figuratively, but there are today numerous online sites displaying what are called "digital attack maps" on which you can, literally, watch "lies" traveling halfway around the world in a microsecond. These maps show, in real time, what are known as distributed denial-of-service (DDoS) attacks. Perhaps because it is the easiest form of cyberattack, it is also the most pervasive. Here's how Keith Alexander, former NSA director, explained a DDoS attack to me: "You remember your kids in the backseat yelling so you and your wife couldn't talk? That's a distributed denial-of-service attack. That can be done with them not knowing much about your facility, only throwing packets of data at you. Overloading the system so that you can't conduct business. If you're a stockbroker or a company that makes its living on the network, that's a huge problem."

A glance at one of those digital attack maps online will complete the mental image: gushing fountains of colored dots, each representing a separate attack, cascading to and from various parts of the world. Many of the lines, as one might expect, emanate from China or Russia and are aimed at the United States. Reverse traffic on the cyber highway is equally dense. Some attacks, however, originate from totally unexpected sources, aimed at equally unexpected targets—Luxembourg to Peru, Russia to Belize, Denmark to the United States. But tracking cyberattacks is not simply a matter of tracing an arc of colored dots. An attack that appears to originate in Denmark may actually have been routed through Bulgaria from a computer in Russia. Even confirming a Russian point of origin is insufficient to sustain a charge that the Kremlin had any knowledge or played any part. And as the level of sophistication rises among individual actors, so too does the degree of deniability across the board.

In the first days of September 2014 a graphic video showing the beheading of American journalist Steven J. Sotloff by a hooded executioner representing the Islamic State of Iraq and Syria (ISIS) was posted on the Internet and was instantly available around the world. Whatever benign intentions the Web's earliest designers may have had, their gift of universal distribution and access came without a filter. ISIS stage-managed Steven Sotloff's execution to horrify and to outrage, and very likely with the intent of provoking the U.S. government into an ill-considered response. The decision by almost all international news agencies not to reproduce the video may have been a reflection of good taste, but it was essentially irrelevant. Once posted, the video of the murder was available to anyone who wanted to watch it.

Over that same Labor Day weekend, Apple's iCloud servers were hacked, resulting in the leak of private photographs of various celebrities. The very term "iCloud" is an interesting piece of marketing misdirection (the cloud's servers are unambiguously grounded) designed to convey the impression that vast amounts of digital information are floating serenely and securely "up there" somewhere, out of reach to everyone but us. We alone, it is asserted, are the arbiters of access to that material. As of early September 2014 Apple's advertising copy still read "With iCloud, you can share exactly what you want, with exactly whom you want." Exactly? Well, not exactly. We don't need to understand how the Internet functions in order to summon its magic, and in truth, most of us don't.

Neither the Grand Guignol of an ISIS execution, the breathless tabloid coverage produced by unauthorized images of naked movie stars, nor the acceptance of a status quo of ceaseless DDoS attacks comes close to approaching the outer limits of where our Internet dependency may lead. When the advantages of a new technology promise so much and the inherent dangers are not yet fully understood, or seem more relevant to others than to ourselves, it is easy to defer action. But a growing body of evidence suggests we can no longer afford such complacency.

The Internet is, after all, a neutral instrument, wholly reliant on the capabilities and intentions of its users. In one set of hands it is a toy, in another a terrifyingly destructive instrument of war. The very interconnectedness that bestows on this medium the capacity to weave a benign global social network also provides the countless tiny paths of access to growing armies of cyber warriors. Those paths lead to dangerous places. One military cyber specialist spoke with me on the condition of anonymity, as he is still on active duty. "We're terrible at assessing risk if we can't visualize it and if it involves something we don't control. If you go to a beach, there are people who will never go in the water because there are sharks in the water— the *Jaws* thing, right? If I go to a vending machine, I put in my quarters and I don't get my drink and I shake the machine. The probability of the vending machine falling on me is 1 in 110 million. The probability of me dying from a shark attack is 1 in 250 million. We suck at looking at risk."

There is, however, an entire industry dedicated to the objective assessment of risk. The business of insurance, after all, depends on nothing less. The insurance industry calculates the odds of something happening and then puts its money at risk. How likely is a successful attack on a power grid? I decided to consult a credible insurance specialist.

Ajit Jain is the CEO of the Berkshire Hathaway Reinsurance Group. If Jain downplays his importance to the investment juggernaut that is Berkshire Hathaway, his boss, Warren Buffett—the "Oracle of Omaha"—has done nothing to discourage speculation that Jain will someday succeed him. In his 2014 letter to Berkshire shareholders, Buffett credited Jain with the establishment of a $37 billion float. Between the time when a customer pays his premium and when the insurance company has to pay out a claim, the insurance company can invest this "floating" capital toward its own profit margin. Ajit Jain's feat in building that large a float, wrote Buffett, is one "no other insurance CEO has come close to matching."

Jain's success is founded in the business of reinsurance, the market in which insurance is sold to other insurance companies. If, for whatever reason, an insurance company finds itself overexposed or overextended, the Berkshire Hathaway Reinsurance Group offers a hedge, a backup, a way of spreading the risk. Like every other kind of insurance, it comes down to studying the record, calculating the odds that an event will occur, and then naming a price to protect the buyer against the financial consequences of that event.

The risks can seem astronomical, but when premiums are accumulated from many sources over many years, the occasional massive payout can be just the price of doing business, as Jain explained: "If a force-five storm or hurricane were to make a direct hit on Miami, the industry is equipped to lose as much as two hundred billion dollars. It would not be a great day, but life would go on." Jain estimated Berkshire's share of that payout at anywhere from $2 billion to $7 billion.

Unlike hurricane insurance, the business of cyber insurance is relatively new. Jain asked a colleague, Kevin Kalinich, to join our conversation. Kalinich is global practice leader for cyber insurance for Aon Risk Solutions, a top U.K.-based risk consultant and insurance broker. It is a business, Kalinich explained, intended to address new risks arising from "the Internet of things," a field in which predicting the likelihood of events is all but impossible. It barely begins to define the challenge of insuring a power company against the cost of a catastrophic cyberattack. Kalinich agreed. "There are certain industries such as utilities, power, electric, water, that have unique exposures." "Unique exposure" refers to the extraordinarily high risk of insuring against new, unfamiliar, and potentially catastrophic events. The industry is still plotting its own cautious road map toward coverage for those exposures; it will, Kalinich told me, require a combination of traditional insurance, the involvement of a reinsurance company such as Berkshire, and government guarantees of limited liability. No one, in other words, can be expected to provide complete coverage. Even the combination of those three elements, according to Kalinich, wouldn't go much beyond $1 billion of insurance to cover a cyber-related event.

That struck me as a pretty trivial amount. Here was Ajit Jain, after all, contemplating an enormous industry-wide payout in the wake of a massive hurricane striking Miami head-on, with what could be described as a $200 billion shrug. The impact of a successful cyberattack on a power grid could be far worse, I suggested, and Jain didn't disagree. "If there were a complete blackout in a certain part of the country for a three-month stretch," he said, "the looting and everything that [could] ensue just boggles the mind, how large the numbers [would] be."

At this stage, Jain contended, it would be premature for anyone in the insurance industry to talk seriously about "calculating the odds" of a catastrophic cyberattack. There's not enough history, he said, not enough data points on which to base those calculations. Though insurance companies can anticipate and, with some accuracy, even calculate the cost of a hurricane, insurance for cyber-related events is uncharted territory.

"Having said that," Jain continued, "we all have some subjective notion of something like that happening; we slap [on] a margin of safety and set aside a certain amount of money." While the sums he's talking about are not exactly play money, Jain said that there is only so much that companies such as his are willing to risk in an "emerging field" like cybersecurity. Jain has not become a potential successor to Warren Buffett by taking unsustainable risks. "The[se] extreme scenarios . . . are certainly likely, and we can all debate how likely and what do we mean between likely versus unlikely. But from the insurance industry's perspective, the amount of exposure that we are willing to take on is nowhere close to the exposure that would come from these very extreme events." With so much still unknown about the risks involved, his company may soon offer insurance against cyberattacks, but it will likely demand such a high premium that there will be few buyers.

Insurance is an unsentimental business. It is based on the proposition that a certain number of customers who buy insurance to protect themselves financially against one misfortune or another will have to be paid in full. As long as those payments are significantly less than the cumulative premiums paid in, the company stays profitable. For the time being, and given the capacity of the electric power industry to protect itself or to reconstitute itself after an attack, the consequences of cyber sabotage against a power grid are too uncertain and potentially too enormous to merit the establishment of a realistic business model. Without government support and the guarantee of limited liability, it appears that the insurance industry is prepared to dip a toe in the market, but not much beyond that.

Are we any closer, then, to determining the likelihood of a massive attack against the grid? Only to this extent: the insurance industry won't bet against it.

Preparing the Battlefield

Proposed *Jeopardy* question: "A quadrillion
operations a second." Answer: "What is
a petaflop?"

Any successful attack combines three features: opportunity, capability, and motive. As we have seen, even the most ardent defenders of grid security acknowledge its vulnerability to cyber intrusion. There is disagreement as to how much damage could be inflicted, but the arguments over opportunity and capability are issues of degree. But what about motive? The ancient Romans posed the question *"Cui bono?"* To whose benefit? We have barely begun to count the number of potential beneficiaries.

During his 2013 State of the Union Address, President Obama tried to focus the American public's attention on the evolving danger of cyberattacks. "We know foreign countries and companies swipe our corporate secrets. Now our enemies are also seeking the ability to sabotage our power grid, our financial institutions, our air traffic control systems. We cannot look back years from now and wonder why we did nothing in the face of real threats to our security and our economy."

Nothing? Really?

The trove of classified information revealed by Edward Snowden should reassure any concerned citizen that, at least in the realm of gathering intelligence, the U.S. government is innocent of all charges of doing nothing. If anything, the Obama administration, like the Bush administration before it, has been at pains to convince the American public that our intelligence agencies are restrained in their activities. In reality, while the United States may have fallen behind in a number of critical areas, accumulating intelligence data is not one of them. Having said that, it is still the case that the United States and its infrastructure remain highly vulnerable to cyberattack, and there is blame enough to go around. In trying to find the proper balance between security and profit, many industries still incline toward a shortsighted emphasis on profit. Information sharing among businesses and between business and government is undermined by the drive to compete and by fears of litigation sparked by concerns about privacy. Civil libertarians, worried about real and potential violations of privacy, are often insufficiently focused on an even greater need to address external threats to liberty. Washington turf battles, in which a variety of agencies struggle for primacy, undermine the greater responsibility of protecting our most vulnerable targets. Meanwhile, we may have fallen victim to a distinctly American tendency to stress size over performance, confusing the accumulation of data with the gathering of actionable intelligence.

In March 2012, writing for *Wired* magazine, former NSA

employee and longtime chronicler James Bamford reported on a project that the NSA had launched in 2004. The project's goal was nothing less than the creation of the world's most powerful supercomputer: a machine that could execute a quadrillion operations a second, a capacity endearingly labeled a "petaflop." This advanced the fastest computer speed then known by a factor of one thousand. Now completed, this computer (located in Oak Ridge, Tennessee, once home to the Manhattan Project) is linked to another NSA facility on the outskirts of Bluffdale, Utah. Innocuously named the Utah Data Center, this is almost certainly the largest data mining and storage center in the world. The combined capabilities of these two operations are more than sufficient to send chills down the spines of privacy advocates and civil libertarians across the country.

Not so many years ago, Washingtonians joked that the acronym NSA stood for No Such Agency. To this day, its heroes and heroines labor, for the most part, in anonymity. Public acknowledgment of any kind is rare. It is worthy of note, then, that "part of one building," as the agency's former chief scientist puts it, at NSA's Oak Ridge facility has been named the George R. Cotter High Performance Center. The man so honored has been in the front ranks of those in the intelligence community struggling to analyze, develop, and understand the nature of cyber warfare. Now retired, Cotter has devoted himself to studying the vulnerabilities of the electric power industry. He is convinced that China and Russia have already penetrated the U.S. power grid, both for purposes of reconnaissance and, very likely, in order to plant cyber weapons that could be activated at some time in the future. Having spent a lifetime in the intelligence community, Cotter cannot help but arrive at certain conclusions based on what he learned over the years. He insists that none of the material he uses nowadays is classified. Still, when he discusses his assumptions about the Chinese and the Russians penetrating the U.S. power grids, he is also implying what he knows about U.S. cyber activities. Standard operating procedure for any major nation-state, he told me, is to "study the vulnerabilities. You develop attacks against those vulnerabilities. You may actually insert the attack in the system. The general term in military parlance is 'preparation of the battlefield.' That is, you're all ready to push the 'go' button if you have to." We can assume that the NSA, where Cotter spent most of his adult life, has engaged in precisely this sort of "preparation of the battlefield" within the critical infrastructure of America's potential enemies.

The notion that the United States and its principal rivals routinely fire cyber shots across one another's bows also makes sense. Few if any of these cyber skirmishes are acknowledged, but what is publicly known certainly points to the conclusion that they are taking place. Over the course of 2014, rising tensions between Moscow and Washington over events in Ukraine led the United States and its European partners to impose a series of economic sanctions on Russia. Moscow refused to back down, continuing its policy of support for pro-Russian rebels in Ukraine and continuing to apply economic and military pressure on Ukraine's new government, trying to force it back into Russia's orbit. Yet its public response to the U.S. sanctions was surprisingly mild.

That only applies, of course, to what can be directly attributed to Moscow. In August 2014 the *New York Times* reported that a Russian crime ring had "amassed the largest known collection of stolen Internet credentials, including 1.2 billion user name and password combinations and more than 500 million email addresses." In early October of that year, ten American financial institutions were revealed as targets of a huge cyberattack. The most serious intrusion was against JPMorgan Chase, from which the hackers had gained access to the names, addresses, phone numbers, and email addresses of some eighty-three million households and businesses. An anonymous senior official speculated to the *Times* that the attack could have been in retaliation for those U.S. economic sanctions on Moscow. Only days later, reports surfaced that Russian hackers had exploited a vulnerability in Microsoft Windows to gather intelligence on "several Western governments, NATO and the Ukrainian government," as well as "European energy and telecommunications companies and an undisclosed academic organization in the United States."

None of these attacks can be directly attributed to the Russian government, but neither is there any evidence that the Russian government is cracking down on local crime rings targeting the United States. While it is reasonable to assume that Russian criminals might be motivated to harvest hundreds of millions of Internet user names and password combinations with an eye to selling them, why would they be spying on Western governments and NATO? Like the *Times*'s source, George Cotter believes that the rash of cyberattacks on U.S. banks during the summer and fall of 2014 does, in fact, constitute a warning from the Kremlin, related to events in Ukraine—a demonstration to Washington of what might follow if economic sanctions escalated. "Can we prove it? No," he conceded. "[Not] without getting into the heads of the people who are running cybersecurity operations for the Russian intelligence services."

The impact of those cyberattacks on American banks was only enhanced by the fact that whoever launched them could do so behind a cloak of deniability. A skillfully executed cyberattack serves the multiple purposes of inflicting damage and conveying a strong warning, all the while permitting the attacker to deflect accusations with a posture of innocent indignation. One can only speculate on the impact that such uncertainty might have had on U.S. policy makers weighing the option of further economic sanctions against Russia for its activities in Ukraine.

In Cotter's estimation, the calibrated application of such cyber blackmail has been under way for some years, and its users extend beyond Moscow and Beijing. It would go a long way toward explaining the on-again, off-again nature of U.S. foreign policy toward Syria. According to Cotter, Syrian leader Bashar al-Assad "has a cyber operation which he routinely runs against Wall Street," intended as a strong message to the U.S. government. These attacks tend to be relatively low-tech distributed denial-of-service attacks against American banks, but, Cotter suggests, Assad "is demonstrating that if you unleash an attack against the Syrian armed forces, against the Syrian government, all hell will break loose" in the United States' financial sector. Cotter believes that Syria's sabotage capabilities have unquestionably restrained our government's actions against Assad. "I don't think that signal is misunderstood by the White House. I don't think it's misunderstood by any thinking person who understands how the cybersecurity game works."

Former NSA director Keith Alexander wouldn't go so far as to link Syrian cyberattacks on Wall Street to any wavering in U.S. policy toward Damascus, but he did caution against dismissing Syria's cyber capabilities. Those, he pointed out, have been greatly enhanced by Iran, which has dispatched its own experts to train and assist its Syrian allies.

Iran continues to demonstrate its sophistication with widespread and near-constant applications of harassment in cyberspace. A U.S. security firm, Crowdstrike, spent much of 2014 tracking a group of Iranian hackers. They found that the hackers had the potential capability not only to spy on but also to critically damage sensitive networks in the United States, Canada, Israel, India, Qatar, Kuwait, Mexico, Pakistan, Saudi Arabia, Turkey, the United Arab Emirates, Germany, France, England, China, and South Korea. In April 2015 researchers from Norse, a cybersecurity company, and the American Enterprise Institute issued a report concluding that "Iranian hackers are trying to identify computer systems that control infrastructure in the United States, such as the electrical grid, presumably with an eye toward damaging those systems." Among the key points in the report was that hundreds of thousands of domains registered to Iranian citizens or companies are hosted by companies in the United States, Canada, and Europe and are then used to conduct cyberattacks on America and its allies.

Countries such as Iran and North Korea (of which more in the next chapter) cannot hope to match the United States in their ability to project conventional or nuclear force; what the Pentagon describes as "kinetic power." Cyberattacks, however, present second- and even third-tier military powers with a fresh avenue for projecting force in the heartland of their enemies, all while enjoying that additional element of deniability.

It requires no great feat of imagination to construct a scenario in which Tehran would deploy the most damaging weapon available against the United States. The Israelis, particularly under the leadership of Prime Minister Benjamin Netanyahu, have left little doubt as to their intentions should Iran cross the nuclear threshold. An Israeli air attack on Iran's nuclear complex might be carried out even over adamant U.S. objections, and although there is no way to foresee Tehran's reaction, it seems reasonable to predict that Iran would believe the United States to be complicit. Iran lags far behind the United States in the development of intercontinental ballistic missile systems. Neither its navy nor its air force would be any match for its U.S. counterparts. There are, however, at least two ways in which Iran could project a retaliatory strike against the United States itself: terrorism or cyber war.

Certainly the United States, Russia, and China are keenly aware of how much damage an all-out cyber war would generate on all sides. It seems thus far to be encouraging a measure of restraint, at least in relation to the power grid. But this awareness has not inhibited the development of cyber weapons, and at this stage there is no realistic way of assessing how any one of the great powers would respond in the wake of a major cyberattack, especially one whose point of origin might not be quickly or reliably determinable.

It all comes back to issues of vulnerability, access, and motive. Several cyber specialists in the military were willing to talk to me, but only on the basis of nonattribution. The Internet, one of them explained, began in a climate of innocence, useful for nothing more than "sharing professors' good ideas." The older systems are vulnerable because they were designed without any thought that they might come under attack. The people who designed them are themselves older, their expertise no longer relevant to the technology being built today. But if one were looking for a way to exploit that old technology, much of which is still in use, one of those older technicians could be just the person to map the route of attack. It's a point worth considering in the context of well-funded terrorist groups looking to acquire instant cyber expertise.

George Cotter told me that the Russian government has created a civilian corps of "largely criminal elements," an all-star team of hackers useful because their government "can simply deny responsibility for anything that occurs," as with the 2014 bank attacks. This is hardly the first time in the history of warfare that "sleeper" agents have been planted under cover within the homeland of a potential enemy. Arguably, though, none has ever been equipped with a weapons system as versatile, as potentially destructive, and as easy to deploy as the Internet.

Military and intelligence experts with whom I've spoken are unanimous in asserting U.S. superiority in launching cyberattacks. There is similar agreement that the nation's cyber defense capabilities are more modest. It's a function, many believe, of operating within the constraints of a democracy. Mike Hayden, former head of the NSA and director of central intelligence, described the handicap as being denied "home field advantage." He was referring specifically to the Fourth Amendment privileges that protect the privacy of U.S. citizens. There is no denying a certain wistfulness among senior American military and intelligence officials when they discuss the constraints of the Constitution.

8

....

Independent Actors

The bad guys are awfully good.
—GEORGE R. COTTER

Until May 2012 Howard Schmidt was President Obama's White House advisor on cybersecurity. What would he say, I asked, if the president asked him directly, "Howard, is there a way we can guarantee that a cyberattack won't knock out one of our power grids?"

"Absolutely not," said Schmidt.

When I spoke to him in the summer of 2014, he and Tom Ridge, the first secretary of homeland security, were partners in a cybersecurity consulting firm in Washington, D.C. Schmidt confirmed what other specialists had been telling me: the greatest cyber threats to the U.S. infrastructure are in the hands of the Russians and the Chinese. Schmidt also echoed the assumption that China and Russia, encumbered by a network of interlocking interests with the United States, would likely be constrained from launching a full-scale cyberattack on an American power grid. Could they do it? Yes. Would they? Only in the context of an expanding crisis.

As one moves down the capability scale of potential actors, though, a disturbing phenomenon becomes apparent. Iran, for example, presents somewhat less of a threat than China or Russia in terms of its capability but has far fewer overlapping interests with the United States. North Korea is yet several notches below Iran on the capability scale but has almost no interlocking interests with the United States and therefore even fewer restraints. In some ways most worrisome of all is the realm of individual hackers, whether independent or at least not visibly associated with a national government.

When I asked Schmidt about independent actors, unrestrained by a network of interests with the United States, he focused initially on profit-oriented groups: "Independent actors and independent terrorist groups that contract to each other— we've seen coordination among those guys stealing the financial data, sharing it with another group that wants to send spam and collect all. There is much more organization from independents than there was in the past." That's bad news for American businesses and the U.S. economy, but it doesn't quite rise to the level of a strategic threat against the United States.

What about independent actors using cyberattacks to knock out one of our power grids? Are we at that point yet?

"Simple answer," said Schmidt, "yes. And that worries me as much as a nation-state using an aggressive move for whatever reason."

George Cotter also sees a growing level of sophistication among criminal organizations, terrorist groups, and so-called hacktivists (political activists who use the Internet, such as the group Anonymous). "I believe," said Cotter, "it is literally possible for a hacktivist group, well trained and well motivated, to take down major portions of the grid without the industry being able to stop it."

In a rapidly changing world, we are obliged to consider certain harsh realities. Whatever conditions may constrain some nation-states from launching a genuinely crippling cyberattack do not apply to an outlaw state such as North Korea or to a growing number of criminal or terrorist organizations. What distinguishes the terrorist organizations from the nation-states can be summarized in two words: *goals* and *consequences*. The actions of nation-states, unless and until they are at war, do not have a simple goal of destruction. Even if the Russian government is behind some of the anonymous cyberattacks targeting the United States, its motivations likely run the gamut from intelligence gathering to sending a warning signal. As we've seen, nation-states are restrained by an understanding of networked interests and likely consequences.

An independent actor such as Al Qaeda or ISIS, in contrast, has as its immediate goals inflicting pain and instilling terror. These groups are, if anything, trying to provoke violent reaction from their enemies. To the degree that such groups have been inhibited from using weapons of mass destruction, it has been due to the difficulties in acquiring and deploying them. Cyber warfare employs a wide-open battlefield with multiple points of vulnerability, an easily accessible weapons system, and legions of experts available for hire. ISIS, for example, has the money (it is believed to have accumulated more than $2 billion in assets), and it has the motive. It is not yet clear whether it has found the experts. But in the opinion of the NSA's former chief scientist George Cotter, "if ISIS can recruit a trainable, competent cyber army, then what they need to develop is available for a price in the black market."

General Lloyd Austin III is the commander of United States Central Command (CENTCOM), responsible for the twenty-country area of responsibility (AOR) consisting of Iraq, Syria, Afghanistan, Pakistan, Iran, Egypt, Lebanon, Yemen, Jordan, Qatar, Kuwait, Bahrain, Saudi Arabia, the United Arab Emirates, Kazakhstan, Kyrgyzstan, Oman, Tajikistan, Turkmenistan, and Uzbekistan. If there is a likely breeding ground for a terrorist attack against the United States, it can be found somewhere on CENTCOM's operational map. Austin cites the growing divide between the Shia and Sunni branches of Islam, the tensions between moderate and radical Islamists, and the "youth bulge," the group of educated, unemployed, and disenfranchised young people who are prime candidates for recruitment by extremist organizations. These, fueled by widespread anti-American and anti-Western sentiments, constitute a foundation to the growing threat against vital U.S. interests. It is precisely among young, educated radicals, warns Austin, that a new generation of cyber warriors will be recruited.

And then there is North Korea, which straddles categories: a nation-state with the instincts of a terrorist organization. Sources within the U.S. military's Cyber Command told me that the North Koreans, while less advanced than the Iranians, are well along in their development of cyber war capabilities, due in no small part to instruction by the Chinese and Russians.

The volatile mixture of a rogue state, uniquely isolated, with an unpredictable leadership emerged in bizarre fashion during the waning months of 2014. The cyberattack on Sony Pictures Entertainment became an awkward case study of America's commitment to the First Amendment. It will, in time, be seen as a dangerous escalation in cyber warfare—recognized as such at the time by President Obama, but widely misunderstood by a distracted public.

A Hollywood film, *The Interview*, applied a broad brush of locker room humor to the most isolated nation on earth and its brutal and unpredictable ruler, Kim Jong Un. In what appeared to be an act of retaliation directed by North Korea, hackers took the film's production company, Sony Pictures Entertainment, offline. That's a benign way of saying that Sony was publicly humiliated and, for a period of months, had its corporate computer system rendered inoperable. The hackers dumped onto the Internet five Sony films that were due for first-run theater release. Privileged information—executive and superstar compensation packages, medical records, budgets—was made public, as was a trove of silly texts and emails that perhaps shouldn't have been sent in the first place. That was, the hackers suggested, merely an appetizer. They claimed to have a hundred terabytes of Sony data. In what may have been the most unheeded advice since Lot's wife ventured a final glance back at Sodom, Michael Lynton, Sony Pictures Entertainment's CEO, urged his employees not to read the waves of pirated emails, because it would cause them to turn on one another, damaging relationships.

The North Korean government denied any involvement in the hacking attack, even as someone speaking in the voice of the attackers warned of dire, 9/11-type consequences if Sony actually released *The Interview*. That was enough for the four largest theater chains in the United States, which announced that they would not carry the film. Sony initially followed suit, canceling release of *The Interview*, only to be roundly condemned by some of Hollywood's biggest stars and, more significantly, by the president of the United States himself for failing to uphold freedom of speech. During his final news conference of 2014, President Obama expressed disappointment that Sony executives had not sought his opinion. "We cannot," said the president, "have a dictator imposing censorship in the U.S."

The *New York Times* and the *Wall Street Journal* ran lengthy investigations, citing the FBI and laying the blame squarely on North Korea. The *Times* suggested that digital techniques were used to steal the credentials and passwords from a systems administrator who had maximum access to Sony's computer systems. The *Journal*'s version reported that while Sony had installed no fewer than forty-two specialized computers designed to keep hackers out, in another example of human error trumping technology one of these so-called firewalls apparently went unmonitored when Sony shifted from an outside company to an in-house team.

With the ultimate release of *The Interview* in about three hundred theaters and news that a digital blackout had effectively, if temporarily, wiped out North Korean access to the Internet, it appeared that justice of some sort had been served. The new year dawned, and public appetite for the story waned.

The national import of the attack on Sony is hardly in the past, however. It highlighted a number of threats and vulnerabilities that had already preoccupied Obama's national security team and which are destined to plague future administrations. As damaging as the cyberattack on Sony may have been, it never constituted an obvious threat to national security. If, however, a skilled team of hackers can disrupt a large corporation in the entertainment field, what's to prevent them from launching equally devastating attacks on American infrastructure? It is not that Sony was unprepared or unprotected. The forty-two firewalls cited by the *Wall Street Journal* were designed to protect the company against precisely the sort of attack that took down its computer system. This raises questions about the vulnerability of smaller, less profitable corporations with fewer resources than Sony (or JP Morgan Chase, or Target) to spend on cybersecurity. It's a point that former NSA director Keith Alexander emphasizes with particular reference to the hundreds of electric power companies that are simply unable to afford the best cybersecurity, while remaining connected to the same grid as the companies that can.

President Obama also recognized the significance of this attack in particular. While he addressed the issue in the context of supporting a film's right to be seen, his message went far beyond defending freedom of speech. The president took what appeared to be an isolated assault on a private corporation and raised it to the level of an attack on the national interest. This, he made clear, was not merely cyber theft, nor was it intelligence gathering. What had been inflicted on Sony Pictures was an attempt at cyber blackmail. In pledging that the United States would, at a time of its choosing, "respond proportionately" against North Korea, Obama struck the posture of a leader sending an unambiguous message to his counterparts around the world. North Korea had come dangerously close to crossing a red line. Just as U.S. policy precludes paying ransom to or negotiating with terrorists holding an American citizen hostage, the president was defining America's position toward blackmail carried out in cyberspace: no deals, guaranteed retaliation. After all, what had been applied to an entertainment company could be deployed against vital U.S. interests.

Warning of consequences, though, is one thing; delivery is another. What is most dangerous about Pyongyang and its mercurial leadership is not only its unpredictability but also its degree of immunity to cyberattack. North Korea has so much less to lose in a high-stakes cyber war than the cyber-dependent United States; it is neither easy nor particularly effective to isolate a hermit kingdom. The concern that President Obama expressed a few years back to his aides about the vulnerability of the U.S. infrastructure and America's dependence on computer systems applies inversely to North Korea. As of late 2014, the total number of Internet protocol addresses in North Korea was estimated at 1,024—"fewer than many city blocks in New York," the *New York Times* observed. Satellite photographs of the Korean peninsula at night speak volumes: South Korea is ablaze in light, while north of the demilitarized zone at Panmunjom is a nation almost entirely plunged in darkness. There is not much infrastructure to target in North Korea.

Major General Brett Williams, the former director of operations for United States Cyber Command, dismissed President Obama's warnings in the wake of the Sony attack as "empty statements." Going further, he declared that "there's nothing we could do in cyber. Certainly nothing we are willing to do, because it crosses a lot of red lines."

Indeed, when administration officials briefed reporters on

retaliatory measures that had been taken against North Korea, they turned out to be "largely symbolic" economic sanctions levied against ten North Korean officials and an internal intelligence agency.

While cutting off the Internet connectivity of a country that scarcely connects with the outside world may be less than a body blow, prudence imposes limits on other forms of retaliation. North Korea has artillery in abundance—thirteen thousand pieces, by South Korean estimates. Deployed mostly along the demilitarized zone separating North Korea from South Korea, the longest-range artillery pieces are capable of hitting Seoul, and as Williams reminded me, the United States has hardly been more effective an enforcer in the physical realm than in cyberspace. "We can't stop them [the North Koreans] from launching missiles over Japan," he said. The North Koreans have, in fact, repeatedly defied warnings from the United States, Japan, and South Korea, launching test flights of their Taepodong-2 missiles over northern Japan and into the Pacific Ocean. North Korea also has—lest we forget—a growing arsenal of nuclear warheads, and the capacity and inclination to share its nuclear technology with America's enemies.

Nations at war have spent much of the past hundred years developing and then trying to preclude the use of certain classes of weapons. They have, with varying degrees of success, employed and then been horrified by the dreadful impact of poison gas, chemical weapons, and of course the atomic bomb. For more than seventy years now, the unambiguous consequences of nuclear warfare have convinced the bitterest of enemies, in the most dangerous confrontations, not to draw their nuclear swords. That knowledge has restrained the United States and the Soviet Union, the Soviets and the Chinese, the Chinese and India, India and Pakistan.

When the United States and Israel collaboratively launched their cyberattack on Iran's nuclear program, they set an entire arsenal of new forces into motion. In its limited goal of delaying the development of a nuclear bomb, Stuxnet apparently succeeded. But in her definitive book on the attack, *Countdown to Zero Day*, Kim Zetter offers a chilling assessment from the executive director of the *Bulletin of the Atomic Scientists*, Kennette Benedict. "We have come to know how nuclear weapons can destroy societies and human civilization. We have not yet begun to understand how cyber warfare might destroy our way of life," Benedict noted. "How ironic that the first acknowledged military use of cyber warfare is ostensibly to prevent the spread of nuclear weapons. A new age of mass destruction will begin in an effort to close a chapter from the first age of mass destruction."

What Benedict described as "the first age of mass destruction" has remained, following the 1945 U.S. bombing of Hiroshima and Nagasaki, in a suspended state of potential rather than actual horror. There have been no subsequent atomic or nuclear attacks. The use of cyber weapons remains in an early stage, and retired U.S. general David Petraeus contended that the development of such weapons has outpaced strategic thinking. Unlike the earliest years of the atomic age, Petraeus told me, when strategic thinkers such as Herman Kahn, Albert Wohlstetter, and Bernard Brodie began developing their theories on nuclear deterrence, we really don't have something similar for cyberspace. These days Petraeus does his strategic thinking for a private equity firm, serving as chairman of KKR Global Institute. The former CIA director argued that thinking on cyber war remains relatively immature for a couple of reasons: "One is that the development of cyberspace has just been so rapid, so fast, that theorists can't even keep up intellectually. Second, whereas really only two blocs of countries had nuclear weapons for a long period of time, that's not true of the cyber arena."

It is at this stage almost impossible to catalogue the identities and locations of the world's most sophisticated hackers. Whatever limited reassurance we have that nuclear weapons remain under rational control does not apply to the use of cyber weapons.

I return to General Lloyd Austin, commander of CENTCOM. Name a crisis or trouble spot in the world today, and more likely than not it falls under Austin's purview. During our interview, I asked him directly, is there a danger that a cyberattack will someday take down a major section of the U.S. electric grid? "It's not a question of if," he said, "it's a question of when someone will try that."

Why, I wondered, does there seem to be such limited awareness of the impending danger? Austin's answer was simple. "We've not experienced a significant effective attack against our power grid or against our transportation networks. So, like 9/11, I don't think people realize how vulnerable you are until they see something happen."

He added, "I think some of the key folks in the banking industry, in the transportation industry, they have clearly realized that there are vulnerabilities that we need to guard against or protect, and they're doing some things about them. But as you connect one system to the other across this nation, there's just a lot of points of entry, a lot of points of potential failure that I don't think people have thought through adequately. The average person doesn't think this kind of thing can affect their lives, quite frankly."

As we will see in the coming chapters, the American public are not the only ones unwilling to contemplate, much less cope with, the eventuality of a debilitating cyberattack against our power grid. The government agencies and civic organizations charged with enabling the nation to recover from catastrophe are also woefully unprepared.

Keith Alexander's many years in the military provide some understanding of those confronting multiple crises simultaneously. He puts it this way: "Everybody's out there fighting today's alligators, and we're talking about future alligators, and they say, 'Look, I've got this problem with ISIS, I've got this problem with Afghanistan, Gaza keeps coming up, I got this wingnut in North Korea; and you're talking about a potential problem.' There's no malice aforethought."

In fact, there's not much aforethought at all.

Part II

A NATION UNPREPARED

9

....

Step Up, Step Down

Oh, I'm sure FEMA has the capability to bring in
backup transformers.

—JEH JOHNSON, SECRETARY OF THE DEPARTMENT OF

HOMELAND SECURITY

Let's be practical. If at this point you remain unconvinced that a
successful cyberattack on the electrical power grid is likely, the
government's failure to adequately prepare for dealing with the
aftermath will seem like nothing less than prudent economy. If,
on the other hand, you're afflicted by nothing more than clear-
eyed skepticism, please stick around. There is a line between
prudent economy and misplaced confidence that warrants critical examination.

The nature of the electric power industry is such that it combines modern technology with antiquated equipment. Some of
that equipment is so large, so expensive, and so difficult to replace that it constitutes an entire category of vulnerability. If
there is one piece of hardware that deserves to be singled out

as critical to the nationwide transmission of electricity, it is the large power transformer.

In order for electricity to move over great distances at maximum efficiency, its voltage has to be cranked up. That function is performed by step-up transformers, which take electricity from a generating station and send it flowing at high voltage along the massive power lines that stretch across the American landscape. At times of peak flow those lines along our roads, railways, and highways can actually be seen to sag under the load of surging electricity. At the end stage of the transmission system, a sequence of step-down transformers does what their name suggests, readjusting voltage to a low enough level that the electricity can be safely delivered to the consumer.

No country in the world has a larger base of installed large power transformers than the United States, and that base is aging. The Department of Energy reports that these "critical component[s] of the bulk transmission grid" are, on average, thirty-eight to forty years old. A senior DOE official told me that age in itself is not of great concern, as transformers have no moving parts. Still, the DOE's own report, published in April 2014, lays out the potential consequences of failure bluntly: "Power transformers have long been a concern for the U.S. Electricity Sector. The failure of a single unit could result in temporary service interruption and considerable revenue loss, as well as incur replacement and other collateral costs. Should several of these units fail at the same time, it will be challenging to replace."

"Challenging to replace" doesn't fully capture the scale and complexity of the problem. To begin, the number of large power transformers (LPTs) in use in the United States is staggering.

There is a great deal of information that the power industry refuses to make public, and the exact number of LPTs is one such statistic. The industry has competitive, security, and antitrust justifications for its reluctance to share data; that reluctance, however, extends to sharing with the appropriate federal agencies. The Department of Energy can only hazard a guess as to how many large power transformers are in use, but it reports that the number "could be in the range of tens of thousands." Even if that number is wildly exaggerated (and it's difficult to imagine why the DOE would do so), the central issue is the difficulty of replacing transformers. "An LPT is a large, custom-built piece of equipment," the DOE report explained. "Because LPTs are very expensive [$3 million to $10 million each] and tailored to customers' specifications, they are usually neither interchangeable with each other nor produced for extensive spare inventories."

Consider the sum of all those factors. Conservatively, there are thousands of aging transformers, most custom-built, unable to be ordered from a catalogue or mass-produced, each costing somewhere in the neighborhood of $3 million to $10 million. Add to this that there are only a handful of plants in the United States capable of building an LPT—as of this writing, ten such facilities. The vast majority of large power transformers are built overseas, and more than 75 percent of those purchased by the U.S. energy sector must be procured overseas. The estimated lead time, the time from production through shipping to delivery, is commonly between one and two years, and never less than six months.

These transformers are so enormous—anywhere from 400,000 to 600,000 pounds—that they cannot be transported on a standard railroad freight car. It requires the use of a specialized railroad freight car known as a Schnabel. There are only about thirty of these in North America, and as one senior FEMA official conceded, some of the original transformers were delivered so many years ago that the rail lines on which they were transported no longer exist. When LPTs are transported by road it calls for a modular device seventy feet long with twelve axles and 190 wheels. The unit occupies two lanes of traffic and requires special permits from each state through which the transport will pass. Because of the enormous dimensions and weight involved, these special permits often call for the prior inspection of various bridges and other pieces of infrastructure along the way.

All of this combines to present a critical liability for the resilience of the United States power grid. In October 2014 I raised the issue with Jeh Johnson, who during the writing of this book was secretary of homeland security and the cabinet official most directly responsible for the security of the nation's infrastructure. I asked the secretary what would happen in the event that several transformers were knocked out. How would he go about replacing them? What kind of a backlog exists?

"I'm sure FEMA has the capability to bring in backup transformers," Johnson said. "If you want an inventory and a number, I couldn't give you that."

I had spent part of an afternoon at FEMA headquarters only a few weeks earlier talking with the agency's administrator, Craig Fugate, about the same issue. Contrary to Johnson's assurances, Fugate had pinpointed the failure of large transformers as one of his greatest areas of concern: "If you cause overloads and you cause significant damage to the very large transformers, that's probably one of the most difficult things" for us to respond to, "because there [are] very few manufactured in the United States." Asked what he would say were Jeh Johnson or the president to ask whether FEMA was prepared for such a scenario, Fugate responded bluntly, "No." He continued, "Most people expect . . . that somehow we have enough tools in the tool chest to get power turned back on quickly. The answer is no."

Notwithstanding Johnson's reassurance that FEMA has the capability to "bring in" backup transformers, there is almost no such capability in the realm of large power transformers. Those, Fugate went on to explain, depend on the power industry's willingness to invest in redundancy. Fugate argued that the industry was more inclined to invest in excess capacity and maintenance personnel before deregulation. In the current climate of greater competition, and with management under pressure to return as much profit as possible to shareholders, the bottom line has taken priority over resiliency, especially among smaller and midsized companies.

Johnson, a lawyer by training, had a number of pressing issues on his agenda the week that we talked: the Ebola virus had just made its initial appearance in the United States, ISIS was threatening U.S. citizens and interests, and there had been a couple of major breaches of security at the White House, reflecting poorly on the Secret Service. I took note of the fact that Johnson must have faced a tremendous learning curve getting up to speed on the range of threats to the U.S. infrastructure, and I asked him to tell me how he had learned about the threats to the power grid in particular. Johnson's answer ran slightly more than thirteen minutes, and he never addressed the question. It was, he conceded a little later, not an area about which he has any expertise. That was why, he said, he had asked several colleagues to join the conversation. He was accompanied by his assistant secretary for public affairs, Tanya Bradsher; by Caitlin Durkovich, assistant secretary for infrastructure protection; and by Suzanne Spaulding, undersecretary for the National Protection and Programs Directorate. In theory, the presence of so many senior officials should multiply the amount of available information. In reality, it has an inhibiting effect.

Durkovich picked up on the question of spare large power transformers. The industry, she said, has a planning scenario, the Spare Transformer Equipment Program (STEP), in which companies would have spare transformers available for their substations. They would also have the ability to lend those transformers to other substations with similar designs. She estimated that there are already between two hundred and three hundred high-voltage supertransformers available. "We are also working to ensure that if they were needed, we could move them across interstate lines in a rapid fashion." She did not address the issue of how equipment weighing half a million pounds or more would be transported "across interstate lines in a rapid fashion."

Executives from the Edison Electric Institute claim that there are, in fact, hundreds of large power transformers available to be used as spares; as with the number of LPTs in use, though, the actual number is "proprietary." Since large power transformers are custom-built, the likelihood that a matching spare can be found for an LPT that fails is small. A couple of hundred spares of differing size and format might provide a patch but not a solution. Spaulding acknowledged as much: "Not every one of those transformers is going to be a plug-in where they might be needed at a substation. There's a recognition that this is an area of concern. And the industry is trying to figure out can they move away from this . . . everybody's different way of operating, so that they can have more interchangeability."

It's easy to leave a layperson in the dust, but Jim Fama of the Edison Electric Institute did his best to explain the issue to me. Transformers come in different voltage classes. A 230–500 kV transformer is not interchangeable with a 345–500 kV transformer, for example. But within a given class, Fama assured me, adjustments can be made, and that, he said, is what STEP is about.

No one is suggesting that there is anything approaching a quick fix. Moving those enormous pieces of equipment represents a massive obstacle no matter what. Scott Aaronson, Edison Electric's director of national security, mentioned another component to STEP, something called the recovery transformer. Think of it, he explained, as the equivalent of one of those small spare tires that will at least get you to the next gas station. One senior FEMA official, speaking on the condition of anonymity, told me the program was being developed by Homeland Security's Science and Technology Division. When it's up and running, the program will daisy-chain two or more mini-transformers as replacements for large power transformers that fail. That program, he said, was being jointly conducted and funded by government and industry, though it is still in the testing stage and nothing has been produced commercially yet.

General Keith Alexander expressed far greater pessimism about the timeline for recovery from an attack on a power grid, estimating that it would take several months. Why so long, when repairing (for example) Hurricane Sandy's damage to the grid took only a week? "Because," he said, "there are parts of the infrastructure that would go down that are not easily replaced, like the transformers. So point number one: get me some more transformers."

"Current waiting time," I offered, "over a year."

"Right. That's probably not acceptable. So you say, 'OK, who's going to fix that? You got that one? How many do we need? Go get it.' Does that make sense?"

Asked why we haven't made planning for this scenario more of a priority, he chuckled. "You're talking about a potential problem. How do I plan for future problems while I conduct current operations? In the military you have the current ops officer, but you also have the plans officer. The point that needs to be made is that this is an issue. It's got to be planned for."

Planning is under way. The largest, most profitable power companies are investing in backup equipment, but as with network security, the industry as a whole remains highly vulnerable. Experimental solutions are being tested, but they're still years away from being available across the board. Large power transformers remain vulnerable to cyberattack, and because of their size and because so many of them are out in the open, they are also vulnerable to a well-trained team of saboteurs armed only with semiautomatic rifles, as was demonstrated in California in 2013. What remains unchanged is that LPTs are essential to the functioning of the grid. Because they are very expensive, only the largest and most profitable power companies can afford to keep backup transformers on hand. Because the transformers are custom-made, they are not easily interchangeable. Because the equipment is huge, it is not easily transported. Because these transformers are, on average, thirty-eight to forty years old, some of them were originally delivered by rail systems that no longer exist. Because the vast majority of LPTs are built overseas, it takes a very long time to replace them. When, as you will see in the next chapter, government officials compare an attack on the power grid to a natural disaster such as a hurricane or a blizzard, keep those distinctions in mind.

10

....

Extra Batteries

We are not a preemptive democracy. We are
a reactive one.

—TOM RIDGE, FIRST SECRETARY OF HOMELAND SECURITY

The existential nightmare that haunted members of George W. Bush's administration during the last months of 2001 and throughout many of the months that followed was an image of 9/11 on steroids. What if some terrorist group was to get its hands on a "dirty bomb," a small nuclear weapon? What if the next attack involved chemical or biological weapons? Almost every action and authorization that flowed from Washington was predicated on one or another of those what-ifs.

Judgments rendered in recent years on the wars in Afghanistan and Iraq, on the black sites and brutal interrogations, on the expanded powers of the National Security Agency and the CIA, the drones and the aerial assassinations executed by remote control, have emerged in the more reflective environment that the passage of time permits. Many of the decisions reached in the wake of 9/11 were wrong, but they were issued by men and women far more frightened by the prospect of having done too little than by the consequences of having done too much.

On September 22, 2001, President Bush asked Tom Ridge, the former governor of Pennsylvania, to join him at the White House as special assistant to the president for homeland security. With the passage of the Homeland Security Act in November 2002, Tom Ridge became the first secretary of the Department of Homeland Security (DHS), responsible for overseeing what had been twenty-two separate departments and agencies as diverse as the Secret Service, the Coast Guard, and the Immigration and Naturalization Service. The department's mandate, conceived as it was through the prism of what had just happened, essentially boiled down to preventing another terrorist attack. During subsequent years, the department's mission has evolved, in the public mind, to what might almost be described as a policy of "protect almost everything against almost anything." Ridge acknowledged that the department ought to be doing even more on at least one level: ensuring security, he agrees, involves not only preventing disaster but also planning for its potential consequences. That part of the mission, Ridge told me, is almost doomed to fail. "We are not a preemptive democracy. We are a reactive one. Rare are the occasions on which we act in anticipation of a potential problem."

There have been, as of this writing, only four secretaries of homeland security. Each of them has conceded the likelihood of a catastrophic cyberattack affecting the power grid; none has developed a plan designed to deal with the aftermath. I met the first, Ridge, in the offices of Ridge-Schmidt Cyber LLC, a

Washington consultancy company in the field of cybersecurity. The Schmidt half of the partnership is Howard A. Schmidt, the former cybersecurity coordinator for the Obama administration, whom we heard from in Chapter 8. He is the technical expert in this duo.

Neither Ridge nor Schmidt, who had served at the White House more recently, was aware of any plan in the event that a cyberattack knocked out our electricity. I suggested to Tom Ridge, who brings a long career of political experience and connections to the table, that most Americans would likely expect the government to have a plan, a way to take care of the public during such a catastrophe. "Correct," he said. "I'm sure they would say [that]. It would be helpful if the political world would just accept that there are two permanent conditions that are going to affect future generations: one is the global scourge of terrorism, the other is the digital forevermore." Within that world of the "digital forevermore" lies the prospect of a catastrophic cyberattack on one of the U.S. power grids. Where, then, might a concerned citizen find advice on how to cope with the aftermath of such an attack?

"There is no answer," said Schmidt. No government agency has guidelines for private citizens because, according to Schmidt, there's nothing any individual can do to prepare. "We're so interconnected," he said, that in terms of disaster preparation "it's not just me anymore: it's me and my neighbors and where I get my electricity from. There's nothing I can do that can protect me if the rest of the system falters." It's an answer bordering on the fatalistic: the individual can't do anything and the government won't do anything.

In the immediate aftermath of 9/11, Ridge tried to engage

the public in confronting the prospect of another terrorist attack. "We can be afraid, or we can be ready," he told a Red Cross
gathering at the time. But while Ridge intended to provide some
guidance, this became an exercise in humiliation. Ridge recommended that people stock their homes with plastic sheeting and
duct tape in the event of a chemical attack, a proposal that made
him the butt of numerous late-night monologues. "Oh, yeah. I
remember well," he laughed. "It's going to be in my obituary."
Ridge's example and the humiliation he endured cannot have
encouraged any of his successors to invest either time or effort
in leading further campaigns in disaster preparedness.

Tom Ridge's successor as secretary of homeland security
during the Bush administration was Michael Chertoff. He estimated that a concerted cyberattack could knock one or more
power grids offline for several weeks. When I asked whether he
believed the American people are prepared for anything like
that, he stated the obvious: "In some parts of the country, people do stock food and buy generators. In urban centers people
don't do that. In New York you'd try to move a lot of people out
over a period of time."

"Really?" I asked. "More than eight million people? Where?
How?"

Chertoff acknowledged that dealing with grid-wide outages
would present a unique problem in public management. "Get
a hand-crank radio," was the former DHS secretary's principal recommendation. It was evident that he hadn't delved very
deeply into the issue. "In a dense city," Chertoff offered as something of an afterthought, "people can walk to schools and fire
stations."

What, I wondered, could I learn from the senior officer at

my local fire department? The captain on duty at the Potomac, Maryland, fire station assured me that there are secret locations where food and water have been stored. "For all of us?" I asked.

"No," he acknowledged. "Just for the first responders."

"What about the rest of us?"

He considered the question for a moment and then conceded that he would be awaiting further instructions.

"And when you get those instructions," I wondered, "how will you communicate them to the rest of us when the electricity's out?"

"I'm due to retire in a couple of years," said the captain. "I'm hoping it doesn't happen before then."

To date, the longest-serving secretary of homeland security has been Janet Napolitano, who put in almost five years on the job before taking the post of president at the University of California. In October 2012 and during the weeks that followed, Napolitano presided over the coordination of a federal response to Hurricane Sandy. In addition to hitting major sections of New Jersey and Long Island, Sandy flooded New York City streets, tunnels, and subways, effectively cutting off all electric power to Lower Manhattan.

"It was," Napolitano recalled, "very cold. It was still wet, so the plan was to mass resources to restore the power. So we brought in, for example, power trucks, flown in from places as far away as California on DOD [Department of Defense] planes, to begin replacing the poles and the lines. At one point FEMA had about eighteen thousand people working in that area going door-to-door, bringing people food and removing them from unsafe buildings until we could get the power back on." Janet Napolitano's recollection was that power to Lower Manhattan was restored within twenty-four hours and that within ten to twelve days power had been restored to the entire region. It actually took more than five days before any power was restored to Lower Manhattan, but 95 percent of New York's customers did have their power back after thirteen days. Even so, thousands of homes were lost throughout the region and tens of thousands were rendered homeless.

But what, I wondered, if a blackout was the result of a cyberattack? What if the affected area covered several states and efforts to restore power were ineffective for weeks or even months? Is there a plan?

"There is no plan that would be adequate in that circumstance," Napolitano conceded. She insisted, though, that the experts are good at figuring out quick workarounds. "One thing we need to do better," said Napolitano, "is to make sure that that capability is firmly entrenched in every power company in the United States. The big ones have it; it's the small ones, the investor-owned utilities, that you really have to worry about." It's an ongoing refrain: whether in the area of security or resilience, the smaller, less profitable companies are the weak links in the vast chain of companies that make up the grids. There remains a void between defining the problem and proposing a solution.

We met Jeh Johnson, current secretary of the Department of Homeland Security, in the previous chapter. Asked to define the threat of cyberattack, he said simply that "it is potentially very large. It is potentially devastating." Could he, I wondered, be a little more specific?

"Well," he said, "the cybersecurity experts could be more specific than me, but I would say, given the interconnectivity of cyberspace, an effective attack has the potential to cause wide-ranging devastation on our power grid, on critical infrastructure." At this point he deflected the question to Assistant Secretary Caitlin Durkovich ("She's smarter than me"). Durkovich insisted that the cyber threat has been overdrawn. She repeatedly stressed the resilience of the electric grid, echoing the power company executives we met in Part I. She credits owners and operators with having put into place mechanisms and redundancies that would mitigate any cyberattack.

It is not surprising, then, that when I raised the issue of a national preparedness plan, something that might give the public some advance notice of how the federal government intends to deal with the consequences of a successful cyberattack on a power grid, the conversation became prickly. Johnson began by enumerating the various federal departments and agencies that would be involved in a response: the Department of Defense, the Department of Energy, the National Guard, FEMA. "First thing I would have to do as secretary is look to our federal emergency response experts in FEMA."

What Johnson had done, I pointed out, was name the component parts of the various agencies he would assemble. "Is there," I wondered, "a plan that you're familiar with for what you would do?"

"Well, like I've said, I would know who to go to, to get my plan on a moment's notice."

"And that's not something you think you might need to know about beforehand?"

"Look, Ted, I hope this is not a memory quiz. I'm sure that in the course of my intensive briefings coming into office I have been given awareness of a contingency plan in the event of a large-scale loss of power. It may even be sitting up there among those white books."

It was not an unreasonable position. Johnson heads the third-largest federal department, behind only the Department of Defense and the Veterans Administration. He has a staff of approximately 240,000. He cannot be expected to be intimately familiar with every problem confronting his department. Still, Johnson was not shy about appearing on the Sunday talk shows in February 2015 to reassure the public after an ISIS threat to attack shopping malls in the United States. When a rumor surfaced that ISIS might use the Ebola virus against the United States, Johnson addressed the threat publicly. The secretary had already acknowledged the potential impact of a cyberattack on a power grid as "devastating." But he was clearly unfamiliar with what his department might do in the wake of such an attack. In any event, it was worth giving the issue another try, I thought.

"Just help me understand a little bit," I asked. "I'm sitting at home. My power has gone out. The water's not working. The toilet's not flushing and the refrigerator and the freezer are starting to leak. Where and how am I getting my information? From whom? And what are they communicating?"

Tanya Bradsher, the assistant secretary for public affairs, began to answer, but Johnson had clearly had enough.

"Wait, wait, wait, wait! Hold on before you answer that! At some point, this is on the individual member of the public to do a little bit to plan for that contingency. So if in your hypothetical you lose power completely—I mean, this is not modern learning here. This is 1960s learning. You lost power completely, you ought to have, you gotta have a radio. That's not a revelation."

The revelation, I pointed out, would be the duration and scope of the crisis. We're talking about something that could theoretically affect millions of people over a period of several weeks.

"But in the immediate aftermath," Johnson countered.

"Fine," I said. "I'm OK for two or three days."

"Maybe my nineteen-year-old wouldn't know how to do this, but I could, in the dark, find my battery-powered radio so I could find a radio station somewhere that works."

We had squared the circle. It seemed pointless to ask just what information would be conveyed on the radio. There is a clear reluctance to accept the proposition that a cyberattack disabling all or part of a power grid is any different from a blizzard. "We had Snowmageddon," said Bradsher, referring to a blizzard in February 2010, "and we got through it. We got through Snowmageddon."

Suzanne Spaulding, the undersecretary, tried to rescue the conversation, acknowledging that, depending on the scale, a cyberattack could lead to an extended, very challenging period. She enumerated some of the steps taken in the wake of Superstorm Sandy: bringing in generators and fuel, setting up shelters to provide heat and power. "Again," she conceded, "it would be a significant challenge."

I ventured one final approach to the secretary, who was clearly ready to end the interview: is there not some value to getting a message out beforehand on what to do beyond the first two or three days?

"I suspect there is a message that is out," he said. "It's just very few people are actually paying attention to it."

When Caitlin Durkovich joined the conversation again, it

was clear that we had been engaging in a dialogue of the deaf. The scenario I was describing, she told me, has a relatively low probability of occurrence. Not to worry.

In successive State of the Union addresses President Obama has warned of the danger of cyberattacks on our infrastructure. Government is adapting to the "new normal" of daily hacking, and cyber specialists such as Richard Clarke and George Cotter, who held senior government posts, have explained that the Russians and the Chinese are almost certainly inside the grid, mapping its vulnerabilities. Keith Alexander and Howard Schmidt warn that independent actors will soon have the capability to damage the grid, if they don't have it already. If nothing else, the United States demonstrated with Stuxnet what a carefully planned cyberattack can do to the most securely defended equipment. Still, senior officials at the Department of Homeland Security, including the current secretary, treat the likelihood of a crippling attack on one of the nation's power grids as nothing more than a speculative threat, and an unlikely one at that. We are, as Tom Ridge put it, a reactive society.

There are plans. Of course there are plans—dozens of them, possibly hundreds.

As we've seen, for all the warnings from within the government and from high-ranking members of the military and intelligence establishments, and despite the known vulnerabilities of the transformers critical to the viability of the grid, there remains a determination within the power industry and among some government officials to stress the grid's resilience. They invariably cite as evidence the manner in which electric power has been restored in the wake of one natural disaster after another. Absent a crippling example to the contrary, the presumed consequences of a cyberattack on a power grid are bundled into the same general category as blizzards, floods, hurricanes, and earthquakes.

On one level, this is understandable and even prudent. Experience is a more compelling instructor than speculation. Indeed, negative experience, such as that accumulated by the Federal Emergency Management Agency during the aftermath of Hurricane Katrina in New Orleans, can be especially instructive. FEMA is a far better-led organization today than it was in 2005. That's the good news. FEMA is, after all, the agency within the Department of Homeland Security that will bear the heaviest and most immediate burden of recovery, no matter what happens or why. A cyberattack may be different from anything FEMA has previously dealt with, but it is not unreasonable for the agency to focus on the experience it has gained from natural disasters.

This approach falters, however, when relevant federal agencies fail to provide for (or in some cases even contemplate) the difference in magnitude between the effects on the grid of any recorded natural disaster and the potential effects of a massive cyberattack. For one thing, the affected area could be much greater. Even the partial blackout of a grid could leave half a dozen or more states without electricity. Also, unless one credits the Old Testament–style intervention of an angry deity, storms do not deliberately target a system's critical weaknesses. Cyberattacks do, and if we assume that the attackers are predisposed to inflict maximum damage, they will try to conceal what they are doing. Stuxnet succeeded in spinning those Iranian centrifuges into a self-destructive mode over an extended period of time, precisely because Iranian engineers were misled into believing that everything was functioning normally, even as the damage was being inflicted.

The associate administrator for response and recovery at FEMA came to the agency from the Coast Guard, from which he retired with the rank of rear admiral. When we talked in September 2014, Joe Nimmich was reluctant to accept my premise of a wide-ranging, weeks-long electric power outage affecting millions of people. Still, if it did happen, he insisted, the federal government would be ready to deal with it. He was confident that electric power sufficient to avoid a catastrophe could be restored quickly. "I've planned for a million people being homeless, I've planned for tens of thousands of people being deceased. I think very easily we can convert those plans." Nimmich was describing a scenario in which Southern California is hit by a catastrophic earthquake. "When we look at the plan . . . we're talking about activating seventy thousand troops." He referenced Title X, the legal basis for the roles and missions of the armed forces, saying that he had planned for "the National Guard to keep law and order, and the Title X forces to be able to go in and actually help people move." Relocation was central to Nimmich's plan. "The plan is, you start moving people east. You take them out of Los Angeles, put them in hotel rooms in Nevada."

A cursory online check revealed 124,270 hotel rooms throughout Nevada. Assuming that they could all be emptied out before the evacuees were brought in, this would suggest about eight people per room. Granted, that is a quibble. In the face of such a catastrophe, people would open their homes, convention centers and basketball arenas would be adapted, and hundreds of thousands of refugees would be transported to other states. Somehow, shelter would be found.

The aftermath of a massive earthquake, though, bears very few similarities to the loss of a power grid to cyberattack. Where FEMA's presumed 9.0 earthquake would leave a city in rubble, with thousands of dead and injured, even the most massive cyberattack would inflict very little immediate physical damage. Following a serious earthquake, the need for evacuation would be unambiguous. Even buildings that appeared undamaged and infrastructure that had not been destroyed could be severely compromised. There would be the constant danger of further collapses. Sheltering in place would not be an option. Returning to the devastated region could be a matter of years.

On the other hand, in the case of a power grid going down, urging people to stay in their homes may be exactly the right thing to do, at least in the immediate aftermath. Buildings would be essentially undamaged and bridges, roads, and tunnels untouched, leaving routes open for resupply convoys and voluntary evacuation for those who choose to leave. There would be the immediate crises of people injured in the unaccustomed darkness and patients suddenly deprived of life-supporting equipment, but none of these emergencies would be alleviated by mandatory evacuation, especially if neither the duration nor the scale of the electrical outage was known. What I was describing to Nimmich was, in terms of immediate impact, far less than that of an earthquake, but potentially extending over a far greater geographic area and involving many millions more.

I put the center of this hypothetical disaster in Manhattan.

Nimmich was undeterred. "If, in fact, for some reason this is going to be a long duration, we are going to start an orderly movement of people out of Manhattan. And whether you bring buses in or you use trains, you're going to have to move them out of the area. You know, you're giving me two alternatives: we either find some way to restore the power or we move people to where they're no longer in a life-threatening situation."

"You're going to move five or six million people?"

"Sure."

Spoken with the confidence of a rear admiral. To Nimmich, there is no clear answer nor is there a specific plan, and there is no plan, he patiently explained, because "the dire straits you have articulated [are] not what we have gotten from the experts that we work with." Which is yet another way of saying, "We haven't planned for it, because we don't think it's going to happen."

Joe Nimmich's boss is the administrator of FEMA, Craig Fugate. Far from being a skeptic, Fugate believes that "large regions of the United States could go dark" in the very possible event of a cyberattack against the grid. I quoted Fugate in an earlier chapter on the vulnerability and unavailability of those large power transformers so critical to the grid.

As for his deputy's mass evacuation plan for Manhattan, Fugate was dismissive.

"Can't move 'em fast enough," he told me.

"You can't move that many people that fast," I echoed, "and anyway, where are you going to move them?"

"Yep," said Fugate. The very agencies that would bear responsibility for dealing with the aftermath of a cyberattack on the grid have yet to find common ground on even the most fundamental questions.

In Washington, where plain answers to blunt questions are a rarity among people still in government service, an interview with Craig Fugate is a refreshing change. His former boss, the previous secretary of homeland security, Janet Napolitano, described him as a particularly focused individual. "Craig," she said, "has only two interests: University of Florida football and disaster relief."

What, I asked Fugate, would he say if President Obama came to him and wanted to know the plan in the event of a prolonged and widespread power outage?

"We're not a country that can go without power for a long period of time without loss of life. Our systems, from water treatment to hospitals to traffic control to all these things that we expect every day, our ability to operate without electricity is minimal." The FEMA administrator expressed a frustration likely common among senior government bureaucrats: "I've got to deal with the consequences" despite not "really hav[ing] any say on the front end as to why we got in this situation."

It's worth noting that when I interviewed Craig Fugate we were alone in his office. In Washington these days, that is a rarity. Most senior government officials are so worried about the consequences of what they say appearing in public that they like to have at least a public affairs officer present during an interview, to modify or mitigate any controversial answers. Not Craig Fugate (or, to be fair, Joe Nimmich).

So what, I asked FEMA's administrator, is the plan for a prolonged, widespread power outage? For the first couple of days, he explained, the primary burden would be on state and local governments, but if the electricity remained out for weeks or more, it would be FEMA trying to fill in the gaps. "The plan would be to support the states to keep security, to maximize what power we do have to come back online, to look at what it will take to keep food and other critical systems like water systems up and running with generators and fuel. To prioritize where we're going to start rebuilding our economy." Fugate warned that there's a limit to how much FEMA can do, but he's confident in prioritizing certain objectives. "Keep the water on," he said. "That means we need to have enough power to pump, treat, and distribute water through the system. You have to keep the water system up, and you've gotta then focus on the water treatment system. Backing up sewage is just about as bad. Those two pieces will buy you enough time to look at what your alternatives are. Basically, people have to drink water, they have to eat, that waste has to go somewhere, they need medical care, they need a safe environment. There has to be order of law there."

Fugate is not a man to mince words. There is traditional disaster response work, which is about reestablishing normalcy very quickly. Then there is uncharted territory, he acknowledged, "where normalcy [wouldn't] get established quickly. We [would be] trying to hang on and keep as many people [as possible] from dying until the system comes back." That's not the sort of message that would inspire widespread confidence in a concerned public, but it has the ring of authenticity to it.

11

···········

State of Emergency

The basic tools of government are extortion and bribes.
—CRAIG FUGATE, FEMA ADMINISTRATOR

In the event of a regional crisis, the first lines of authority run
through state capitols. It's up to the governor of any given state
to mobilize the National Guard, up to the governor to order an
evacuation, up to the governor to request federal assistance. The
day may come when a cyberattack has such wide-ranging consequences that it will have to be treated as a hostile act against the
United States. It will be, quite literally, an act of war. Until that
time, however, the federal government tends to wait until the
states request assistance.

Governor Andrew Cuomo of New York would likely first
turn to his state commissioner of homeland security and emergency services, who was Jerome Hauer back in 2014, when I
interviewed him; Hauer left the post early in 2015. In the late
1990s, when he served as New York City mayor Rudy Giuliani's

director of emergency management, Hauer had been a frequent guest on my program *Nightline*. He was then, and remains, an outspoken and colorful figure. When he welcomed me to his Third Avenue office in Manhattan on October 24, 2014, he was wearing a shoulder holster holding a Heckler and Koch pistol. On the same date exactly a year earlier, Hauer had unnerved a delegation of Swedish security professionals by using the laser on his pistol as a pointer during a presentation.

Hauer wasted no time in expressing his lack of confidence in the federal government's understanding of the power grids, and his conviction that a cyberattack on a control station would have devastating consequences. "If somebody gets into the network, then the ability to reroute is gone. The ability to actually monitor is gone. The ability to black out a control station leaves them [the power companies] helpless."

"Your job," I said, "is to tell New Yorkers what the plan is if the power goes out throughout the state. What's the plan?"

"Well," said New York State's commissioner of homeland security and emergency services, undermining whatever might be left of Admiral Nimmich's proposal, "we're never going to evacuate New York City. What we'd do is set up shelters for people to basically reside in. One of the biggest problems in a city like New York is the high-rise buildings. When power goes out we have hundreds and hundreds of people stuck in elevators. I can't tell you how many calls the fire department gets during a blackout."

It was Hauer who oversaw New York State's response to Hurricane Sandy. Sandy created great hardship for many people, but Hauer, like Janet Napolitano, said that the scale and duration were manageable. "The federal government was terrific," said Hauer. "They brought in millions of meals. They brought in fuel through the Defense Logistics Agency." There were, Hauer explained, millions of gallons of fuel in underground storage tanks, but gas station operators lacking generators to run their pumps couldn't retrieve the fuel. It's another example, albeit a small one, of business owners choosing profit over resiliency, because those generators can cost $50,000 or more. Following Sandy, needing to get the gas stations up and running again, federal government responders pumped $14 million worth of fuel into stations along "critical routes" and New York State installed generators in the majority of these stations, free of charge.

Donating fuel and generators to key stations during a short-term, localized crisis is one thing; convincing the owners of gas stations around the country to install backup generators in anticipation of a crisis is quite another. It would seem like a no-brainer, a way for owners to ensure that their pumps will function even when the power is out. But where the bottom line is at stake, small business owners are reluctant to make the investment. In such circumstances, Craig Fugate explained, bureaucrats are left with what he called the basic tools of government, "which are extortion and bribes. Either I give you grant dollars to get you to do something you would not otherwise do, or I tax you to change behavior for what you will not otherwise do." (Far from being unique to crisis management, this essentially summarizes the entire tax code: for extortion, substitute taxes; for bribes, substitute tax breaks or incentives.)

Aside from fuel, the other government supply initiative Hauer cited was food—those "millions of meals." The notion of millions of meals can seem confusingly reassuring. New York

City has a population of eight million people. There are nineteen million people throughout the state. The upstate region, said Hauer, would overall be more self-sufficient in a crisis situation: there are hunters who store their food, and they have deep freezers attached to small generators. Even so, there are only so many hunters, so many freezers, and so many generators.

The city, though; how long could that hold up?

Without federal assistance, Hauer said, New York City "could probably last for two days." The City of New York has warehoused millions of MREs, or meals ready-to-eat, but with a population of eight million these are nothing more than a stopgap. Any crisis lasting more than a few days would be a struggle. In the case of something as widespread as a grid outage, Hauer explained, New York couldn't rely on federal assistance, because it would be competing with other states for food. FEMA "only has so many millions" of MREs stockpiled, and the private companies that produce them would be overwhelmed; states would have to "get in line." Hauer also underscored the importance of knowing one's constituency. He recounted how, in the wake of Hurricane Sandy, the Red Cross delivered ham sandwiches into certain Orthodox Jewish communities. "I know from experience they don't have what we need for kosher, which is a big part of the city." Hauer's team ordered and distributed upward of a million ready-to-eat kosher meals. There are more than four hundred thousand Orthodox Jews in New York City. Even at the rate of one kosher meal a day per person, that supply of a million meals would be exhausted in less than three days. Nor does that take into account the city's Muslim population; the more devout, who will only eat food that is considered "halal," will eat food that is kosher. The overall population of Muslims in New York City is over half a million.

The disaster relief industry, at least that segment of it dealing with producing and distributing long-lasting food supplies, has its own operational complexities. Half a country away from the Orthodox Jewish communities of Brooklyn and Lower Manhattan is an unpretentious cinder-block building on the outskirts of Salt Lake City, Utah. It was where I met the owner and founder of the Saratoga Trading Company, Jeff Davis, and its president, Paul Fulton. Their company's retail outlet is The Ready Store; they exude the same cheerful energy as the squirrel mascot over the sign proclaiming FOOD STORAGE—SURVIVAL KITS—MRES—EMERGENCY SUPPLIES on the outside of the building.

We were seated around a small conference table on which Jeff and Paul had put out a sampling of freeze-dried raspberries, blueberries, and pineapple. Davis and Fulton didn't want to get into the financial details of their business, but they claim to be one of the larger companies in the industry. What they were willing to discuss were some of the peculiarities of the business. Disaster, for example, has an immediate, global impact. When the tsunami hit Japan in 2011 and the Fukushima nuclear plant was melting down, the immediate impact was such, Davis told me, that even their U.S. clients had to wait three months for their orders.

"And the reason we were at three months," Jeff continued, was not "because we couldn't produce it fast enough; I couldn't get raw goods from the suppliers.... It was a problem with everyone." The same market forces were at work across the industry. There is a limit to how much fresh food is available for processing at any given time. The manufacturers who supply government relief agencies with MREs were having to wait every bit as long for product. They just couldn't get the necessary raw materials.

I had previously explained the premise of this book to Davis and Fulton, and Davis was close to speechless. "Oh, my gosh! . . . That kind of thing is so far beyond . . . The numbers would just . . . It would bury us within days."

Why not just build up the MRE stockpile when supplies are available? It's an issue of shelf life, which for MREs is only five years. "So," explained Fulton, "you look at the MRE manufacturers who are trying to build inventory post the tsunami in Japan; they overbuilt, because when buying of MREs stops, it stops and it stops fast. Having a surplus of MREs means a warehouse full of product that loses value with each passing year. Everybody wants fresh inventory that will last a full five years. So there's no incentive for the MRE manufacturers to build up a massive backlog." The only reason to stockpile would be if they knew for sure that an "emergency will happen in the next five years."

"Good luck," I said.

"Good luck," agreed Fulton.

FEMA and other government relief agencies are in the same boat as manufacturers. Ideally, they would want to buy MREs on short notice, but the industry is incapable of meeting crisis-level demands. Loading up on inventory is another option, but the government is disinclined to spend large sums on contingency planning when there's no immediate crisis brewing, especially given that five-year expiration date. The critical factor, then, is the supply chain. There is a limit to how much fresh food is available for processing at any given time. It cannot simply be turned on at a moment's notice. Freeze-dried foods are a longer-lasting option than MREs—properly processed and stored, freeze-dried and dehydrated product can last up to twenty-five years—but Paul Fulton estimated that even a bare-minimum supply of such food would cost at least $2,000 a person per year.

If Congress was convinced that at some point the government might need to provide emergency food supplies to, say, thirty million people for a year, it could, for $2,000 a head, provide the basics to keep them alive. Could this be part of a solution? The $60 billion cost is hardly prohibitive when you consider how many lives would be at stake. It would probably take the industry years to accumulate the necessary raw materials, but in theory, at least, it seems a viable option. In its 2008 findings, you may recall, the EMP commission projected far greater numbers than thirty million at risk. What can be projected with some confidence is that any crisis—whether EMP or cyberattack—that knocked out electricity for more than a couple of weeks over a multistate area would exhaust emergency food supplies in a matter of days.

Ray Kelly served in the New York City Police Department for a total of forty-seven years, twice as commissioner, from 1992 to 1994, and then again from 2002 to 2013. What concerns him is the proliferation of guns and what would happen to the most vulnerable members of society in a city like New York in the face of an extended crisis and prolonged shortages. "People certainly have the potential for trying to take things by force. What happens in an elderly community, where they're certainly susceptible to being attacked in terms of taking what they have, the limited resources, least able to defend themselves?"

Government agencies at almost every level try to anticipate problems by holding what are sometimes called "tabletop exercises" or "war games." Ray Kelly participated in many such exercises. He cannot remember a scenario, though, in which New York City and its surrounding area were assumed to be without electric power for more than five days. Not, he assured me, that he thinks a protracted blackout is unlikely: "There's a real danger here. And I think we just haven't done nearly enough. There's not enough awareness of it, but also government is asleep at the switch."

There were reports in the wake of Hurricane Katrina that as many as two hundred members of the New Orleans police department were under investigation for deserting their posts. The number of police officers ultimately charged was closer to fifty, but the stresses and challenges facing first responders worried about their own families are not difficult to understand. Rudy Giuliani, mayor of New York City during 9/11, thought the New Orleans example was a reflection of poor training and management. "In a good police department," he told me, "a well-run police department, New York, Chicago, Boston, Los Angeles, I think most of your cops and most of your firefighters, if anything, are going to come and volunteer for duty."

Interestingly, his police commissioner, Ray Kelly, who came up through the ranks, starting as a police cadet, didn't have quite that same level of confidence. "The security implications [of getting first responders to work] are huge. You know, they're concerned about their families, they're concerned about their well-being. So over time when you talk about protecting the points of distribution, that all implies that government workers are showing up and do their jobs, and you can't guarantee that over a sustained period of time. Their ultimate concern, like most human beings, will be their families' and their own well-being."

There are individuals whose preparedness planning will get them through the initial days and even weeks when food runs out. FEMA, the National Guard, and branches of the federal government are focused on finding a way to keep water flowing—enough, at least, to keep people alive and to dispose of their waste—but maintaining an adequate flow of food into the cities and keeping the very young, the elderly, and the infirm alive will depend in some measure on the season. Winter, when there is no safe source of heat, would take a particularly heavy toll. In an environment of crowded, hungry, freezing people, each passing day would presumably elevate the potential for violence. It requires a degree of advance planning well beyond whatever exists to deal with the consequences of a natural disaster.

We are inclined, as Tom Ridge observed, to be a reactive society. We apply unimaginable amounts of money toward dealing with the aftermath of crises. The most conservative estimates put the financial cost of the wars in Afghanistan and Iraq at around $1.5 trillion. Most estimates are significantly higher. The Transportation Security Administration, which came into being as a direct consequence of the 9/11 terror attacks, now employs fifty-five thousand people, with an annual budget in excess of $7 billion. Over the course of the past fourteen years TSA has been funded to the tune of somewhere between $90 billion and $100 billion of protection we didn't know we needed before 2001. Nor, it seems, has the money been particularly well spent. In early June 2015, the Department of Homeland Security revealed that its teams of undercover investigators were able to smuggle dummy explosives and weapons through TSA checkpoints at airports around the country in 95 percent of cases.

We tend to come up with funding after disaster strikes.

Press Six If You've Been Affected by a Disaster

They refuse to fill me in because . . . you know, it's
secret squirrel stuff.
— MARTIN KNAPP

Martin Knapp is homeland security coordinator for Park
County, Wyoming. On the inside of the door to Knapp's office in
Cody is a poster with the bold headline HOMELAND SECURITY
above a photograph of four native Americans holding rifles.
Underneath the photograph is a slogan: FIGHTING TERRORISM
SINCE 1492.

Martin Knapp has a sense of humor with an edge that probably plays well in the Cowboy State. As he pours us both a cup
of coffee, I take photographs of a couple of other slogans on the
walls of his office, these starting to crinkle with age: SOME PEO-
PLE ARE ALIVE, said one, SIMPLY BECAUSE IT'S AGAINST THE
LAW TO KILL THEM. And this one: POLITICIANS ARE THE ONLY

PEOPLE IN THE WORLD WHO CREATE PROBLEMS AND THEN
CAMPAIGN AGAINST THEM.

Knapp is a transplant from Mansfield, Ohio; he can still recall his first day in Wyoming, back in 1971. "I drove up South
Park, passed three cars going the other direction. Two of them
waved to me. Now, where I come from, if I waved to somebody
going down the road, I'm likely to get flipped off, even if I know
the person. Here, people are waving and they don't know me.
But that's the way people are around here."

At the age of eighteen Martin Knapp learned the skills of
a horse wrangler while working for room and board. He and
a group of more experienced cowboys would drive a herd into
Grand Teton National Park, let them loose at night to graze,
and round them up again before they scattered at first light. He
eventually acquired the tracking and hunting skills required to
become a guide for a local outfitter. After well over forty years
in Wyoming, Knapp believes he's almost accepted as a local.

Wyoming, clearly, is not New York or Ohio. "They take care
of themselves," said Knapp, referring to the locals. "They're out
of power? OK, I got my own generator, I got this, I got that.
We'll make do. We lived for hundreds of years without electricity. We can do it again."

Which is why I've come to Wyoming. There is a culture of
self-reliance in rural America. People will use their Medicare
benefits and cash their Social Security checks in Wyoming as
readily as in New York or California, but in principle there is
an antipathy to dependence and an inclination to keep government at arm's length. Perhaps a more affirmative way of saying
that is to suggest that people here will try to solve their own
problems before turning to any government agency, local, state,

or federal. Is such a state, I wonder, any better equipped to confront a crisis for which the federal government has no specific plan?

As homeland security coordinator for Park County, Martin Knapp worries about what might happen on his turf, and what to do about it if it does. Knapp has considered the prospect of an electric grid going down, but there's been no guidance on the subject from Homeland Security in Washington. "In fact," said Knapp, "that even goes as far down as the state level. When I've called or tried to say, 'Hey, I'm working on something here if this happens. What does the state recommend, or what are you going to do?' they won't tell me."

"Won't or can't tell you?"

"Probably a little bit of both. They refuse to fill me in because they don't want it to get out what we're going to do—what they're going to do. I'll say, 'I thought we're on the same team here.' But that's, you know, it's secret squirrel stuff." Knapp doesn't seem overly concerned about the lack of information from either state or federal government.

He did try to establish a working relationship with the Red Cross, but that was hugely frustrating. Knapp called the woman who heads up the chapter in Big Horn Basin and left a message, then emailed. A week went by. No answer. Tried again. No answer. He called one of his colleagues at the Wyoming division of Homeland Security, who gave him the name and number of the top Red Cross official in the state. Knapp called and left messages, yet nothing happened. Weeks went by, then a couple of months. Nothing. Finally Knapp reached a woman in the official's office. "'Well,' she said, 'he's not here. I'll have him call you.' And I say, 'You know what? Never mind. I've been trying to get ahold of someone with the Red Cross for three months. I want to know what I can expect from the Red Cross if I call them for resources. I can't even get a call back or an answer to an email. I will mark the Red Cross out of my emergency operation plan and I will find other means to deal with this,' and I hung up."

That did it. The phone was ringing off the hook in fifteen minutes. The following summer, when there was a big fire on the North Fork of the Little Laramie River, Knapp brought the Red Cross in to help tourists who had been evacuated. But the experience left a bit of a bad taste. "A couple of things I heard that they had done, the Red Cross had done before, was the guy that was the head of the state came up here, and he was going around trying to get donations while the evacuation was going on. I have issues with that."

The Red Cross is the largest nongovernmental relief organization in the country. Its CEO, Gail McGovern, whose background as a business executive led to a position teaching marketing at Harvard Business School, told an interviewer shortly after taking the Red Cross job in 2008 that the charity "has a brand to die for." And it does. What is also indisputable is the organization's skill at fundraising. One Red Cross website lists no fewer than fifteen different ways in which a donor can contribute. To make that process as easy as humanly possible, the Red Cross has in recent years created a means by which those wishing to make a $10 donation can do so by simply texting the message "REDCROSS" to a five-digit number, which conveniently (and repeatedly) appears on television screens following a disaster.

In late 2014, journalists from Pro Publica and National Public Radio published an article titled "The Red Cross' Secret Disaster." It is a devastating account, depicting an organization more concerned with bolstering its public image and raising funds than with maintaining the actual machinery of disaster relief.

Among the findings: emergency vehicles taken away from relief work and staged as backdrops for press conferences; inadequate food, blankets, and batteries in locations where these were desperately needed; tens of thousands of meals thrown out because no one knew where to find the people who needed them. Citing internal Red Cross reports following Hurricane Isaac in 2012, the Pro Publica investigation found that "Red Cross supervisors ordered dozens of trucks usually deployed to deliver aid to be driven around nearly empty instead, 'just to be seen,' one of the drivers recalls. 'We were sent way down on the Gulf with nothing to give.' An official gave the order to send out 80 trucks and emergency response vehicles—normally full of meals or supplies like diapers, bleach and paper towels—entirely empty or carrying a few snacks. Volunteers 'were told to drive around and look like you're giving disaster relief.' "

Not surprisingly, the article drew an immediate and angry response. Rebutting the charge that "the American Red Cross cares more about its image and reputation than providing service to those in need," the Red Cross replied: "Every year, the Red Cross responds to more than 70,000 disasters, most of which are home fires that never make headlines. If the Red Cross cared more about image and PR than providing services, we wouldn't spend time responding to these silent disasters." There was no response to the charge that the Red Cross cares more about fundraising than disaster relief.

It is important to draw a distinction between the Red Cross's institutional and management problems and the efforts of its volunteers, whose contributions are often selfless, even heroic. I had asked research assistant Katie Paul to see what advice a caller might receive from the Red Cross on how to deal with an extended, widespread power outage. Initially she was left dangling in a phone tree that dead-ended on a recorded message, but a later call to a Red Cross office in California was far more productive. The staffer on the line could not have been more helpful, evaluating the pros and cons of getting a generator, touting the value of a community support network, and directing Katie to the "Red Cross Disaster Safety Library."

One committed and well-informed person can provide genuinely valuable information. Getting substantive help seems to depend more on the luck of the draw, though, than on organization. Research assistant Rachel Baye tried a similar approach to solicit information from the Department of Homeland Security.

> UNKNOWN: "The Department of Homeland Security
> switchboard. How may I direct your call?"
> RACHEL: "Hi, I was wondering if I could get some
> information about emergency preparedness?"
> UNKNOWN: "Yeah, one moment."
> RECORDING: "You have reached the Federal Emergency
> Management Agency. Please hold for the next
> available agent."
> UNKNOWN: "FEMA operator."
> RACHEL: "Hi, I was looking for some information on
> emergency preparedness."
> UNKNOWN: "OK. Hold on for just a moment."

RACHEL: "Thank you." [Music for twenty seconds, then a recorded message] ". . . the following options. If you're a member of the media and have an inquiry, press one. For general public affairs questions, please press two. If you're with a state, local, or tribal government and would like to speak to someone in intergovernmental affairs, press three. To speak with someone in community relations, press four. For our office of international affairs, press five. If you'd like to speak with someone in our office of legislative affairs, press six. If you've been affected by a disaster and would like to register for assistance, or if you've already registered and have questions or would like to check on the status of your application, please hang up now and call 1-800-621-FEMA. That's 1-800-621-FEMA. To speak with an operator, press 0 or simply stay on the line. Once again, thank you for calling FEMA's office of external affairs."

Rachel pressed four, hoping to speak with someone in community relations, and got this recorded message: "Public 44 44 is not available. Record your message after the tone. When you've finished . . ." At this point, Rachel hung up. The thought struck me that anyone who has been "affected by a disaster and would like to register for assistance" may not make it through the first six prompts of that phone message.

The greater issue, though, is a lack of consistency. The search for guidance on disaster relief, when there is no crisis under way, no stress involved, cannot be dependent on the design of a more efficient phone tree. It cannot rely on the individual enthusiasm of a staffer or volunteer operator at the Red Cross or be undermined by the bored indifference of a government worker at the Department of Homeland Security. There is no guidance on guidance.

If not the Red Cross, FEMA, or the Department of Homeland Security, where should the interested citizen turn? What is available online can be pathetically inadequate, boiling down to the customary recommendation for two to three days' worth of food and water, warm clothing, a functioning battery-powered radio, and extra batteries. Disaster preparation recommendations usually include a predetermined plan for where and how the family will meet. Beyond that, citizens are largely adrift, left to find their own solutions.

A few of them have. A lot of others are still searching.

Part III

SURVIVING THE AFTERMATH

13

....

The Ark Builders

It's like the ants and the grasshoppers, and how
they froze to death because they didn't prepare.

—ALAN MATHENY, PREPPER

For the most part, public reaction to the possibility of a massive cyberattack has not even risen to the level of apathy. Apathy suggests the awareness of a problem and the decision not to worry about it. We're not there yet. To the degree that government and its disaster relief operations focus our attention at all, they direct it toward the familiar: natural disasters common to our region, or variations on terrorist attacks that have already occurred. Perhaps by definition, preparation for the unknown requires a generic approach.

There is, in any event, a growing movement around the country based on the assumption that neither government agencies nor private relief organizations can be relied upon in the event of any major disaster. A generation or two ago, they might have been called survivalists, but there was an extreme right-wing aura attached to that term, conjuring images of bunkers built to sustain life against aerial bombardment. While such groups continue to exist, they have been modified and largely displaced by a much larger group for whom ideology is less relevant. "Preppers," perhaps most easily described as "those who prepare," can be found across the political spectrum. They are not necessarily prophets of doom, simply those who want to be ready for the worst. As such, they are accustomed to a measure of mockery; they are, after all, only rarely proved right. Dealing with daily life is complicated enough without trying to anticipate and prepare for the hypothetical, no matter how extreme the catastrophe, no matter how unimpeachable the evidence.

In that sense, at least, today's preppers are direct descendants of one of the Old Testament's most famous prophets. Indeed, it is not unusual for preppers to cite his example. It may be unfashionable to link catastrophic disaster to God's judgment, but how interesting it is that Genesis, that bare-bones account of the very earliest days of existence, has no sooner laid the foundation for our journey into history than it diverts into an account of total annihilation. If nothing else, the story of Noah provides evidence that mankind has always been troubled by an undercurrent of worry that what is at present cannot last. Noah is an everlasting reproach to the cynics who mock the ark builders.

Our notions of time may differ from biblical accounts, but Genesis tells us that with only seven days notice of a flood that would cover the earth to the peaks of its highest mountains, Noah built an ark. Genesis is silent on the matter of where Noah acquired the tools, the wood, and the vast quantities of tar with which he sealed the interior and exterior of his enormous ship. Where details are provided, they stagger the imagination. Noah was six hundred years old when God alerted him to the impending cataclysm (although this was, relatively speaking, the prime of life; his grandfather Methuselah died at the age of 969). The ark was to be three hundred cubits long, fifty cubits wide, and thirty cubits high—440 feet from bow to stern (significantly longer than a football field), more than seventy feet wide, and well over forty feet high. Even taking into account that he had three strapping sons to help him, construction would have been challenging. It was designed to accommodate not only Noah and his family but a virtually inconceivable menagerie of creatures great and small. Not, as my childhood memory misinformed me, merely two of each, but two each of the ritually unclean and seven pairs of all ritually clean creatures. (Those spares, some commentaries on the Old Testament suggest, were intended for the ritual sacrifices that Noah would perform in gratitude for God's mercy once the waters receded.)

We are free to speculate on whether some of those spare, ritually clean animals might not have ended up as survival rations. After all, more than a year elapsed between the time the rains began and the day the floods had receded enough for disembarkation, and the Old Testament provides no details on how Noah accomplished the extraordinary task of provisioning his ship for that much time. Rashi, the eleventh-century French rabbi, suggests that the task of building and provisioning the ark actually took 120 years—sufficient time for mankind to mend its wicked ways. That resolves some practical issues while raising others best left to biblical scholars.

I offer the story of Noah simply as evidence that mankind has been struggling with the prospect of impending disaster since the beginning of recorded time and that genuine preparedness is a considerable, perhaps even existential, challenge. But preppers of every era have been outnumbered by the skeptics who tend to view their activities with a combination of fascination and amusement. Unless and until we are actually confronted by disaster, we have a tendency to view it primarily as remote, more applicable to others than to ourselves. Disaster viewed from an appropriate distance can even become entertainment.

Wrapped in the right packaging, doomsday scenarios remain a well-established genre within popular culture: *Blade Runner*, *The Hunger Games*, *Mad Max*. We have a deep-seated fascination with our own annihilation, as long as these postapocalyptic nightmares come with a predetermined running time—a limited experience in a darkened theater from which we can emerge, pulse racing, energized and relieved to be back in a familiar setting.

In October 2013 the National Geographic channel broadcast a docudrama called *American Blackout*. Its setting is a major city in the wake of what appears to have been a cyberattack on the power grid. The program convincingly evokes the mounting hysteria and a dizzying spiral of inflationary madness that reduces a city and its environs to a barter economy in a matter of days. We meet and follow a family of "super-preppers" who have retreated into an armed, camera-equipped bunker encampment. We watch with mounting horror as a *Mad Max* scenario unfolds: armed and desperate neighbors preparing to kill members of the small prepper community for their stored food. Eventually the producers of *American Blackout* resort to the oldest device in theater, the deus ex machina. Without explanation, without further analysis, the script resolves all crises by simply turning the power back on. We have vicariously endured ten horrifying days—and then it's over. One moment the nation is in darkness, the next it is again awash in electric light. You can do that on a television special.

The National Geographic channel has been successfully marketing catastrophes and their management since the premiere of its ongoing series *Doomsday Preppers* in 2011. This is popular material. The show's first program of its second season on the air drew more than a million viewers, making it the highest-rated season premiere in what was then the channel's eleven-year history. *Doomsday Preppers* has a catholic appreciation for disaster. Indeed, it is up to the featured preppers on each episode to elaborate on the nature of the crisis they most fear. A National Geographic production crew then "produces" a simulation of the chosen disaster, "inflicting" it on that episode's voluntary victim(s).

Though the disasters vary from earthquakes to EMP attacks, National Geographic's two-man team of experts (they are co-owners of Practical Preppers, LLC) render judgment on how well the featured preppers in each segment would be able to withstand the disaster they fear. The same standards are applied in each case: How long will food and water supplies last? Can they easily be replenished? How viable are security and shelter plans? What about those elements the show refers to as the "X-factors," things like an ability to stay in touch with one another in all circumstances or to administer first aid? Certain conventions remain nonnegotiable: there must be months' worth, if not years' worth, of potable water and nonperishable food, or, alternatively, the capacity to grow and hunt food.

One episode featured a couple who raised a variety of herbs and vegetables in a greenhouse, providing themselves and a fertile family of rabbits with food. The rabbits did their bit by doing what rabbits do, providing a cycle of nourishment and rabbit fur. Self-defense is a consistent issue. The ever-changing cast of preppers relies, for the most part, on conventional weapons such as knives and guns for security, although one prepper in Augusta, Maine, uses handmade tomahawks and throwing sticks.

There have been other such programs: *Prepper Hillbillies*, *Doomsday Castle*, and *Doomsday Bunkers*. They are all, first and foremost, entertainment. Their actual survival depends on popularity, not gravitas. Of all the challenges facing their featured subjects, few, I suspect, match the brutal environment of television ratings.

The fact remains, however, that absent any guidance from Congress or the executive branch of government, beyond broad recommendations for weathering the first seventy-two hours or so, individual Americans have been left to select their own approaches to the prospect of a lengthy, widespread loss of electric power. Among those who have taken up the challenge, some are serious and well organized and know what they're doing. Most don't.

Making sensible, long-term arrangements for surviving the aftermath of a disaster is not easy. It takes considerable time, effort, and often money. Throughout the following chapters you will meet people who have devoted years, small fortunes, and backbreaking labor to the proposition that disaster, in one form or another, is destined to strike. They consider some kind of systemic breakdown inevitable, and they have little confidence that local, state, or national government will come to their immediate assistance at a time of extreme emergency.

Jay Blevins, author of *Survival and Emergency Preparedness Skills* and the organizer of several prepper expos, estimates that there are somewhere in the neighborhood of three million preppers around the country. The statistic is difficult to confirm, as anyone who lays in an extra case of water and a six-pack of tuna fish is free to consider him- or herself a prepper. The vast and amorphous whole constitutes a ready and growing market for the makers of guns and knives, water filtration systems, and dried foods that will be edible unto the next generation.

Considering the gravity of the different potential disasters that drew these people together, it was an amiable crowd inside the social hall of the John H. Enders Fire Company and Rescue Squad. We were in Berryville, Virginia, a bedroom community about an hour outside Washington, D.C., on a warm Sunday afternoon in early August 2014. The town is still more country than suburb. A statue of a Confederate soldier stands atop a granite pedestal outside the nearby Clarke County courthouse. The pedestal is engraved with a cautious tribute to the cause: "Erected to the memory of the sons of Clarke who gave their lives in defense of the rights of the states and of constitutional government." A nearby plaque makes passing reference to an interim era: "The last public hanging in Clarke County occurred here in 1905."

Approximately six hundred people paid a $5 admission fee to attend the weekend-long EC PREPCON III. That $3,000 went to the Fire Department. Attendees heard lectures on everything from canning vegetables to stories of heroic survival against overwhelming odds. Blevins, one of the speakers and the organizer of this event, recounted half a dozen such inspirational tales, including one of a rock climber whose arm was pinned under a boulder: he cut off his own arm, rappelled down a sixty-five-foot cliff, and then hiked eight miles before receiving assistance. You have to prepare your mind for chaos, Blevins told the crowd. You must be open to the possibility of catastrophe, and you must have the will to survive. What use is all of your gear, he asked rhetorically, if you don't know how to use it?

Between lectures the social hall was abuzz with the good-natured chatter of preppers and curious onlookers drifting among the displays of water filters and Navy SEAL killing knives. A "Beginner's Bug-Out Kit" was tucked between a table with gas masks and a poster advertising "a practical guide to nuclear biological chemical warfare." A sign politely informed potential buyers, "We have taken the liberty of getting you started on the emergency supplies you have been meaning to put together for an abrupt departure into the unknown." These gatherings are agnostic as to the circumstances requiring a person's "abrupt departure into the unknown"; there seems to be a tolerance among preppers that tends to avoid unnecessary acceptance of one catastrophe over another. Nuclear, biological, or chemical attack (referred to knowingly as "NBC") receives the same respect as a solar flare, economic disaster, or earthquake. When I raised the prospect of a cyberattack on the electric power grid, the possibility was treated respectfully, but I could tell people were just being polite. In any case, whatever cataclysm worries you, this bug-out kit claimed to be able to carry you and one other person safely through the first forty-eight to seventy-two hours—for $499.99:

"Our Two Person Beginner's Bug-Out Kit Includes:

- One signal whistle
- One Survival Aid 5 in 1 Tool
- One 20" pocket saw
- One 2 Pack of emergency candles
- One Quick Clot Combat Gauze
- One emergency tinder kit (starts fires without matches)
- Four 2 packs of Warm Pack/Hand Warmers
- Four Glow Sticks
- Two Emergency Blankets
- One Emergency Poncho
- One 12 pack of fire starter sticks
- One emergency drinking water germicidal tablets
- One Gerber Survival Compact Multi-Tool
- Six MRE's (Meals Ready to Eat)
- One folding shovel/pick axe
- One Light Weight emergency tube tent
- One 10'x12' Reinforced Tarp
- One Set of Two Motorola Talkabout two way radios (Up to 23 mile range)
- One Set of 6 snares with Printed Instructions
- One 45 yd. Roll of Duct Tape"

The list speaks (however unintentionally) to the futility of prepackaged "instant preparedness." A flashlight would probably be of more assistance than the glow sticks. A crank-powered radio would be of greater use than the two-way Motorolas. Anything my bug-out partner and I have to say to each other, we would simply yell, one emergency blanket to another.

There is something the marketers of this particular bug-out kit are not sharing with us. They don't expect this emergency to be over in seventy-two hours, and if we need printed instructions with our set of six snares, we're probably not going to make it on our own anyway, once the MREs run out. Bug-out kits are only intended to transition you from wherever all hell is breaking loose to someplace else where the supermarkets may still be open or where you have an adequate stash of long-term supplies. Not to question anyone's motives, but these bug-out kits are not cheap. Nothing on display was cheap; disaster preparedness is a flourishing business. Many of the convention attendees, it seemed, were swept up in the reassuring premise of survival itself, unconcerned how or when they might actually put their purchases to use. Serious preppers, I soon discovered, tend to assemble the contents of their bug-out kits themselves.

Two weeks after EC PREPCON III wrapped up in Berryville, Virginia, the Mid-Atlantic Emergency Preparedness and Survival Expo opened for business at the Washington County Education Center in Boonsboro, Maryland. Among the knives, guns, fire starters, and water filters were freeze-dried foods from the Saratoga Trading Company.

Bill Cirmo runs the prepper-catering Bibo Outfitters, Inc. The priciest item in Cirmo's inventory, at $18,900, is a "bug-out trailer." This is designed, he explained to research assistant Rachel Baye, for a very long-term, high-impact event. He cited a number of examples: a nuclear explosion, an EMP attack, a chemical attack. A prepper himself, Cirmo is a fan of William

Forstchen's novel *One Second After* and is convinced that the book's premise of an EMP attack is a credible threat. Along with full nuclear, biological, and chemical suits and decontamination stations, the trailer carries batteries charged by the trailer's solar panel and wind-powered generator. Cirmo assured Rachel that, equipped as it is with a water filtration and distillation system and a thirty-day supply of food, one could just hook it up and drive it away.

Rachel raised a moral dilemma: the trailer would equip Cirmo to take care of himself, but how would he respond to someone who hadn't prepared, who needed help?

"As far as helping my neighbors," said Cirmo, "there's a limit to that. When you give away supplies, you're endangering yourself. I'm sorry that I prepared but you didn't. It wasn't raining when Noah built the ark. The people who said, 'Why are you building an ark, Noah? There's no rain.'" As noted, among the occasional-prepper crowd, Noah is something of a patron saint. Consider the ark mankind's earliest bug-out trailer.

Alan Matheny was demonstrating how to use a Taser. "Battery-powered. Replaceable battery. Five-year shelf life. Depends on how many times you take it out and practice with it or dry-fire it." It is, he explained, a transitional weapon. You can't carry a gun and you need to get to your car. One step leads to another. Once you reach the car, you'll have access to the gun(s) in the "get-home" bag—"you know, there I might have a couple rifles, a couple thousand rounds of ammo, you know, whatever you've felt comfortable with prepping."

Matheny looks upon prepping as being a matter of personal responsibility. "That's what this country's missing." He found the prospect of depending on government support in a crisis futile. "I'm just going to sit here on my front step and wait for the government to bring me my food in a disaster? It didn't work out very well, did it? You know, everybody's got to take personal responsibility."

What would happen, Rachel asked, if others requested his help? If friends or neighbors who hadn't prepared asked, say, to share his food?

"That's going to get probably ugly at some point," said Matheny. "It's like the ants and the grasshoppers, if you've ever heard that story, and how they froze to death because they didn't prepare. At some point you can't give everything away. We like to help who we can, but at some point you have to take care of number one."

The idea of a prepper's movement is something of an oxymoron. For the most part, these are people who put a great deal of stock in individual responsibility. They anticipate government failure and are innately suspicious of large organizations. At the same time, in some quarters there is the recognition that long-term survival in the face of a widespread catastrophe requires a variety of skills—the establishment of what would amount to emergency communities. The more fully envisioned of these involve intricate professional hierarchies, with an understandable emphasis on practical skills ranging from nursing to plumbing. In this cosmos, doctors and mechanics are demonstrably more useful than lawyers, journalists, and politicians. In subsequent chapters we'll spend time with a community that has taken this type of planning to an extraordinary level.

Such arrangements require more thought and planning than most are ready to invest. Greg Johnson, another prepper at the Mid-Atlantic Preparedness and Survival Expo, said that his plan includes bringing a small community of family and friends together, but he conceded that there's no real organization. There's a vague understanding that everyone will meet at his house, then head down to West Virginia to stay with a friend who has a farm. It's not altogether clear that the friend in West Virginia has signed on to the plan.

Some Men *Are* an Island

It's what God would have done if
he'd had the money.
—GEORGE BERNARD SHAW ON VISITING WILLIAM RANDOLPH
HEARST'S ESTATE AT SAN SIMEON, CALIFORNIA

American history reveres personal enterprise and individual
initiative. It also has a healthy respect for the power of wealth.
Ray Kelly, who has spent a professional lifetime policing New
York City, twice as its police commissioner, has learned to assume over the years that the wealthy tend to find their own solutions. If New York City were to lose electricity for an extended
period, Kelly told me, "the more affluent communities will try
to buy their way out of it." Cradle of creativity that it is, New
York City may be particularly well suited to producing urban
legends. The one passed on to me by a middle-aged executive
in the financial industry may or may not have the additional
virtue of being true but confirms the Kelly theory. It contends
that a wealthy Wall Street family has engaged the services of a

former Navy SEAL who has secreted a boat along the east bank of the Hudson River. In the event of a major disaster, the SEAL will take the family in that small boat to a larger boat, securely anchored offshore. The larger boat will then set sail for a safe harbor somewhere in the Caribbean.

Brian, who is a financial advisor in New York, didn't worry about disaster preparedness at all while he was single, but now that he has a family he has given it some thought. His first inclination is to stay in place and keep the family in their apartment. He has purchased something called a "water bomb," which holds up to a hundred gallons and has a pump at the top (clearly it would need to be filled before the power and water pressure went out). Brian also recommends keeping a safe in the apartment and having lots of cash on hand. He is reluctant to own a gun but has acquired pepper spray and a Taser.

If and when he and his family are no longer able to stay in the apartment, he would arm himself with pepper spray and the Taser, go and get his car, pick up the family, and head for the Hamptons. Brian and I were, on the occasion of our conversation, actually in the Hamptons, and it does seem like a lovely place to hunker down in bad times. Still, I pointed out, if a major part of the Eastern Interconnect is down, the Hamptons would also go dark. Brian has considered this. Several of his friends have equipped their expensive houses with large generators. The conversation trailed off as we considered the likelihood that some of these friends, with limited supplies of food and water, might not choose to take in Brian and his family. Ever the capitalist, Brian considered an arrangement under which he would finance the provisioning of his friend's house, with the understanding that he and his family would be ensured a safe haven in the event of a crisis.

Having the money to spend on disaster preparation is one thing; spending that money prudently is another. It depends, in part, on how worried you are and how much money you have. Apparently the combination is flourishing. According to the Freedonia Group, a Cleveland-based market research firm, spending on residential security has more than doubled since 2001, from $7 billion to a projected $16 billion in 2016. That little nugget was tucked away in a *Wall Street Journal* story on what may be the ultimate domicile for the worried wealthy: a decommissioned missile silo near Concordia, Kansas. Entirely underground and constructed to withstand a direct hit by a nuclear device, it has been reconfigured into a number of luxury survival condominium units. There is a community swimming pool, exercise gym, spa, movie theater, and lounge, and the old missile control center has been turned into hydroponic gardens and an aquaculture system. The units sell for $1.5 to $3 million and each comes with a five-year supply of freeze-dried and dehydrated food. The security guards have special-forces training, and although the silo is connected to the electric grid, there is backup power in the form of two diesel generators, a wind turbine, and a battery bank.

Craig Kephart prefers making his own arrangements. He bears a passing resemblance to Hank Schrader, the DEA agent in the hit television series *Breaking Bad*. Craig is slimmer and fitter but carries himself with a little of the swagger that marks the Schrader character. When he picked me up at Lambert–St. Louis International Airport, Craig had already put in some early road time. He had, he explained, been on his regular Saturday morning bike ride—twenty miles or so while the temperature was merely in the low nineties. Over the next ten hours I came to appreciate how much Craig embodies the aphorism "It ain't bragging if you can do it."

Craig is an avid prepper, and the biking is an essential ingredient of his approach to disaster preparedness. He and his wife, Gayle, each have adult children from previous marriages, but now they live in a large, handsome house in an upscale St. Louis suburb. Craig is executive director of the COPD (Chronic Obstructive Pulmonary Disease) Foundation. Craig's work entails frequent trips to Miami, New York, and Washington, D.C. For all the preparations he's made to survive a variety of disasters, Craig worries that he may be trapped out of town and that all conventional forms of travel could be shut down. He always carries enough cash so that, no matter which city he's in, he would be able to buy a bicycle, biking shoes, and whatever other equipment he would need to take him back to St. Louis.

I was skeptical. "D.C. to St. Louis? How far is it from D.C. to St. Louis?"

"I have no idea," said Craig. "What is it, a thousand miles?"

Craig assumes that he could ride 150 to 200 miles a day. He's thought about this a lot. "Last place I want to be is in a major metropolitan area during a time of national crisis."

Gayle, meanwhile, would be quite literally holding down the fort. The Kepharts' suburban home has two safe rooms for refuge if the house came under attack. A narrow, vertical safe holds a variety of long guns and handguns. Gayle traveled to Las Vegas to take a course titled Target Focused Training.

"What it does," she explained, "is teach you to target certain areas of the body. It's not ninja stuff. It's poke the eyes out."

Not just the eyes.

"If someone had told me that I would be grabbing totally strange men's . . . I would have told you that I would rather die than try that."

"You're talking about grabbing someone's genitals?"

"Yes. Grabbing strange men in the groin. But I completed the course, and I think it was one of the best things I've ever done."

She and Craig both have concealed-carry licenses. The possibility of using that handgun weighs on Gayle as she considers what might happen. "I think about it often, especially with Craig traveling. I sit there on my couch with a weapon on the coffee table in front of me, and I go through the scenarios. What if somebody breaks through that door? Could I get to my gun fast enough? Would I? Could I? It goes through my head constantly, and I guess my only answer is you never really know until you're faced with that situation."

Craig and Gayle have both done more than think about the situation; they've discussed it with their firearms instructors. One of them told Craig: "If you ever pull that gun in a situation, it will cost you $10,000. Whether you're right, whether you're wrong, it doesn't matter. You will get sued civilly. You could be prosecuted."

"OK, then," Craig asked, "what's the rule of thumb?"

As Craig reported, the instructor said, "If you're in a situation and you have time to think, 'This is going to cost me $10,000,' that is not the time you'll use that gun. It's when you are in such fear for your life that the only thing you would do is grab that gun and use it." Craig added, "His point was, it's not a conscious decision."

Prepping in general is, for Craig, very much a conscious decision—a long-term, carefully planned, and deliberate process. "That's one of the problems with prepping," he said. "People buy all this stuff and don't know how to use it. So the skills you obtain are much more important. How do you can and preserve food? What about hunting, fishing? How about sanitation and first aid? If we can't access antibiotics, a scratch or a cut can become life-threatening. There are some guys near here, former Rangers, who offer a survival and first-aid course. The other skills I'm trying to learn—how to build privies."

By almost any standard, the Kepharts seem pretty much prepared for anything. They have several months' worth of food in the basement. "We'll put up a year's worth of chicken and then eat it over the winter." In the event of a major catastrophe, though, Craig decided some time ago that it wouldn't be enough. "Our house is not defensible."

Heading out on I-70 west, Craig drives the two of us toward a turnoff about sixty miles from St. Louis. We are in rural Missouri on what might be described as a Protestant road to salvation. The Warrenton Church of God cautions that "anger is only one letter away from danger." We pass the Church of Christ and the Pentecostal Power Ministries, the Fellowship Baptist Church and a stark, simple cross looming above the no-nonsense sign for the Warrenton Christian Church. Grace Bible Church humbly suggests that we "be imitators of God." We are driving through the Bible Belt toward eighty heavily wooded and isolated acres where, in extremis, Craig and Gayle Kephart and members of their extended family of children and grand-children could survive disaster. Not all members of the family are convinced they would take advantage of the opportunity.

Like many preppers, Craig is indifferent to the precise nature of what might drive him and his family to seek extreme shelter. Echoing Bill Cirmo, he credits his awakening to William Forstchen's *One Second After*, which he said provided him with a keener understanding of how dependent we are on electricity. These days, Craig would more accurately be described as a disciple of the Keith Alexander school of thought. As the former NSA chief put it, if you're hit by a car, it doesn't matter whether it's a red car or a blue car.

Before joining the COPD Foundation, Craig Kephart was a successful business owner. Having cashed out, he is now able to harness money to determination in order to support a conviction. He does not, he told me, want to be the guy who puts a quarter of a million dollars into a bunker. He has, however, invested significantly more than that in what he refers to, using a common prepper phrase, as his "bug-out property." It is a place to live, as distinct from being just a place to hunker down and hide.

There's a forty-by-sixty-foot metal barn erected on poles sunk into concrete footings. It has a high bay that accommodates a large recreational vehicle and which would serve as living quarters for part of the family. The other half of the building is, for the time being, a weekend cabin. The roof faces south so that solar panels will get maximum sun. There's a cesspool. There's a well. Craig is contemplating a second building, something more permanent, with a basement and a root cellar. Craig envisions a vault, a secure place to store weapons and ammunition and for long-term food storage. Ideally, said Craig, this would become a permanent residence, which would entail selling the house in the St. Louis suburbs. When we talked, Gayle was not yet sold on taking that final step.

In the meantime, Craig continues preparing his property for long-term survival. A major element of Craig's long-haul plan was a lake; there wasn't one when he bought the property, so he purchased two additional two-acre parcels of land to accommodate a road and a dam. Then he brought in a lake builder, who excavated three acres of land down to an average depth of about twenty feet. During the excavation they found clay with which to construct the dam, and Craig hired a geologist to confirm that it would be adequate for holding water. Looming above the completed dam are three-hundred-foot limestone bluffs; below are about sixty acres of catchment area where rainfall gets funneled into the lake. Craig has now stocked the lake with fish, which were small and not moving much on the afternoon I visited—perhaps due to the blistering hundred-degree heat—but Craig assured me that they will grow to catchable, edible size.

We drove around the property on an all-terrain vehicle that remains on the site. Craig paid the local electric company to put in the poles and lines and hook up the property, so under normal circumstances there is electricity, but he also has a couple of generators. He showed me where he plans to put in his orchard and where the beehives will be someday. Craig estimated that the lake project alone cost about $50,000 and that, all told, the finished property will cost somewhere in the range of half a million dollars. That seemed on the light side.

Not far from the lake, Craig has carved out a small shooting range. I happily accepted his offer to let me try my hand at target shooting. What made a particular impression on me was the setup. From the back of his car Craig pulled out a folding table, on which he laid out the guns, the ammunition, and a first-aid kit. He was scrupulous about showing me the kit's contents, explaining that if I inadvertently shot him, there might not be time or the opportunity to spell things out. Even more than the unexpected difficulty of placing my shots into a man-sized target from only fifteen feet away, that precaution stayed with me. Craig approaches every undertaking by preparing for the unexpected, and doing so with care.

Craig Kephart is applying what he has—money, determination, and a great deal of time and effort—to sustain him and his family in the aftermath of disaster, in whatever form it comes. It would not be fair to suggest that these are the mindless expenditures of a wealthy dilettante. But what Craig has constructed and assembled in rural Missouri is clearly an enterprise beyond the means of most Americans.

Farther west, outside Cody, Wyoming, another determined survivor has invested an enormous amount of sweat equity instead of money. The last time you ran into someone like Andrew Rose may have been back in junior high school, at the science fair. There were the predictable entries: the combination of Diet Coke and Mentos, the modeling-clay volcano looming over a village assembled from a discarded Monopoly set. Andrew was the kid who compared the effect of electrical load on a fuel cell and rechargeable batteries. (That, actually, was the winner of a more recent seventh-grade science fair, but there have always been science fair winners and the rest of us.)

These days, Andrew Rose works at the University of Wyoming

helping manufacturers with product development and providing energy audits and assessments for businesses that would like to reduce energy use through retrofitting. He, his wife, Stephanie, and their eight-year-old son live about a mile up the road from Wyoming's Heart Mountain Relocation Center, where Japanese Americans were interned more than seventy years ago.

As one approaches the Rose family's hilltop home, the most dominant feature is a sixty-foot mast, atop which sits something that looks like a three-bladed, single-propeller airplane with a very large tail fin. It is, Andrew later explained, a ten-thousand-watt wind turbine, capable of producing an average daily energy output of twenty-two kilowatt-hours. Envision ten 100-watt bulbs casting bright light for an hour and you have the equivalent of one kilowatt-hour. The wind turbine generates twenty-two times that much energy every day.

The Rose family home is on the electric grid but could survive without it. Andrew has installed solar panels and a hydronic heating system. The hot water from a propane-powered boiler flows through coils under the brick-and-tile floors. Andrew's various systems generate more energy than the home actually requires, so he sells the excess back to the power company. Though the company pays him only 50 percent of the power's retail cost, by the end of the year his energy expenditures essentially zero out.

Andrew and Stephanie consider themselves early adopters. "When I hear 'prepper,'" said Andrew, "I think prep school. But for what we've done, people say you're an early adopter because you're one of the pioneers to develop renewable energy on a small scale."

The pioneer label seems apt. They're about fifteen miles

from Cody in one direction and fifteen miles from the next town, Powell, in the other. The nearest neighbor is a quarter mile away and, as Stephanie said, "with an eight-year-old, playdates are a challenge."

Andrew built the house, he said, "because I feel it's an act of being self-reliant. It's using resources on our land. It's creating electricity that's carbon-free. There are absolutely no emissions to the electricity that we generate, and I think it's also a good way to show by example that you can do some things to create your own energy."

Andrew Rose is a role model but not a practical example. He lives in a house that he built from the ground up. The walls are ten to twelve inches thick, made of thousands of adobe-like bricks, and—this probably distinguishes Andrew from all but a handful of other Americans—he made the bricks himself.

"Right, right. I built it with local materials, natural materials. I learned how to make the adobe through a lot of reading. I also did a four-day workshop in adobe and rammed-earth building." The blocks alone, he said, took five to six years to make. "And then, of course, there's a lot of exposed wood, so there's wood beams and a lot of unusual features to the house. There's arches, curved walls, one fireplace built within a wall. A lot of unusual touches that make it unique."

"Andy," I asked, "would you be offended if I told you that a lot of people out there would say you must be some kind of a nut job?"

A man who spends five years assembling thousands of adobe blocks is not easily offended. "I'm a little bit eccentric," he allowed, "and that's OK."

Stephanie offered up her own definition of what it means to

be an early adopter: "Someone who is willing to try something that would not fit into the status quo at that point in time."

What they've done, Andrew added, can serve as an example for the future. The concept, called "distributed generation," is not unique to Andrew Rose. It envisions downsizing the current system of large-scale power plants to clusters of smaller generators spread across a broader area. His own home is an example of how this cottage-industry energy plan might work.

None of this is going to be of any immediate value to a family in Chicago wondering how they will survive the loss of a power grid, but it provides a glimpse into a more sustainable future. In that very real sense, Andrew Rose and his family are pioneers. Movements may grow out of what they are doing. In the meantime, they are neither discouraged by their current isolation nor particularly driven to find converts. Still, many of the advantages enjoyed by millions of people today emerged from designs and inventions produced by innovators inspired by nothing more than insatiable curiosity and an indefatigable drive. It seems important to acknowledge that Andrew Rose and I are made of different stuff. A hundred and fifty years ago, I would have dropped off the wagon train in St. Louis. Andrew would have made it through the Donner Pass.

15

....

Where the Buffalo Roamed

*We lived for hundreds of years without electricity.
We can do it again.*
—MARTIN KNAPP

My first interview of the morning wasn't until ten, and I'd planned on sleeping in. When the phone rang at seven, it was Alan Simpson, the former Wyoming senator and a longtime friend, whom I had asked to put me in touch with a few locals who might offer a uniquely western outlook on surviving disaster. He'd clearly been up for a while. "You had breakfast yet?"

I lied. "I was just heading out."

"I'll meet you in front of the Buffalo Bill Center at eight."

My interest in the Buffalo Bill Center would have been modest at any time of day. At 7:00 a.m. it was nonexistent. But the combination of shame and gratitude is a powerful motivator, and here was Simpson, just shy of his eighty-third birthday, having set me up for every interview I was doing in Cody, full of piss and vinegar and ready to take the time to walk me through this tourist-trap museum of the Wild West.

The Buffalo Bill Center of the West, it turns out, is a gem. We paused briefly at the holographic image of Buffalo Bill Cody welcoming visitors to the museum. "That's my older brother, Pete," said Simpson. He wasn't kidding. Pete Simpson, all done up in western gear, looks as close to a transparent Buffalo Bill as anyone has a right to expect. The hologram's kid brother and I raced through the Buffalo Bill section (what a wonderful old hustler he must have been) and into the Plains Indians section. The Plains Indians, Simpson explained, agreed to be a part of this project only if they had total control over their part of the exhibit. They were fed up, he said, with being displayed through the white man's prism.

We zipped through the Whitney Western Art Museum, with its spectacular Remington bronzes, and lingered for a few minutes in the Cody Firearms Museum. Then Simpson had to be off for an interview—leaving me, as he no doubt intended, mulling over just a few of my misconceptions about Wyoming and the people who live there.

The point of visiting Wyoming was precisely that it is not New York or Los Angeles. It is not only a different environment, it is a different culture. Disaster preparedness is a matter of upbringing and common prudence. The skills that empower self-sufficiency are ingrained from childhood. Guns are a frequently seen and acceptable part of daily life.

I was about to enter a world of contradictions. By happy coincidence, I found myself in Cody at the same time as a number of Japanese Americans who had been interned at the Heart Mountain Relocation Center during World War II, just down the road from where Andrew Rose's house now sits. The site of this camp, which housed 10,700 people of Japanese ancestry between 1942 and 1945, lies about thirteen miles northeast of Cody. There's a memorial park there now, and members of the diminishing pool of survivors still come back every so often to remember. Among those is Norman Mineta, former mayor of San Jose, nine-term Democratic congressman from California, secretary of commerce under President Clinton, and secretary of transportation under President George W. Bush.

Mineta and Alan Simpson have been friends since Boy Scouts, when Simpson's Cody troop would pay occasional visits to the troop of Japanese American boys at the internment camp. On one occasion, when Mineta and Simpson shared a tent, Simpson conceived a scheme to get back at a boy who'd been hassling Norm. Heavy rain had been in the forecast, and the kid's tent was downslope from Simpson and Mineta's. Simpson carefully dug a shallow ditch around their tent, opening up a channel that diverted the rain directly downhill. The rains came and revenge was served, as it should be, cold. Seventy years later a couple of octogenarians chortled as they recalled the cries of outrage.

There's a handsome monument now where the Heart Mountain Relocation Center once stood. Erected in 1985, it declares, "May the injustices of the removal and incarceration of 120,000 persons of Japanese ancestry during World War II, two-thirds of whom were American citizens, never be repeated." It's nice that Park County erected the monument, even nicer that the Heart Mountain High School alumni class of 1947 paid for it, but that was forty years after the end of World War II, with the stern judgment of history to motivate them. What's more difficult to understand is that only a year or two after Pearl Harbor, the parents and mentors of a Boy Scout troop in Cody encouraged their children to befriend members of a Japanese American troop. It gives some credibility to the claims of a tradition of neighborliness and community values that one hears repeatedly in this part of the country.

The first thing you notice about Bob Model, who is in his early seventies, is his striking resemblance to Teddy Roosevelt. It is a resemblance he cultivates. A great-grandson of William Rockefeller, the founder of Standard Oil, Model is still identified on some websites as a New York businessman, but the identity he clearly cherishes is that of a Wyoming rancher and past president of the Boone and Crockett Club—yes, that Boone (Daniel) and that Crockett (Davy). The club was founded in 1887 by Theodore Roosevelt and to this day carries out its founder's mission of promoting conservation and "maintaining the highest ethical standards of fair chase and sportsmanship."

Eastern urbanites like myself are simply unprepared for the scale of a large western ranch. Bob Model picked me up at the Simpson house, dismissing my insistence that I could find my own way, and indeed I would have missed the turnoff from the highway. The ride to the main ranch house is ten miles along a dirt path called Rattlesnake Creek Road. Model's property, Mooncrest Ranch, encompasses 5,000 acres, surrounded by about 175,000 acres of federal forest land. From roughly June 1 through the end of the year, his horses and cattle are free to graze across that enormous expanse. To all intents and purposes, then, when Bob Model considers the natural resources at his disposal he is looking at what exists on about 180,000 acres. If campers and hunters want to pass through or even hunt on his property, Model requires only that they check in with his ranch office. There's plenty of room to go around, and Bob Model is a great believer in western hospitality.

Some of the ranch buildings are on gravity-fed wells, but the main house is supplied with water by electric-powered pumps. Should the power fail, and out here it often does, there are generators and hundreds of gallons of diesel fuel. Blizzards are common in this part of the country, and Rattlesnake Creek Road becomes impassable to motor vehicles. Model estimated that, based on the food they keep at the ranch, they could survive a month. Beyond that, there's 180,000 acres of nature's supermarket.

"If the grid were to disappear on us overnight," he told me over a plate of homemade brownies, "we always have the ability to hunt." There are large herds of deer and elk. "We would be fine. We have plenty of flour on hand, so we'd be able to make bread. We have an ample supply of wood and a woodstove to cook with. Yes, it would be challenging, but this is something we think about. We don't do practice runs, but we do think about it. Living the way [we do] and where, [we] are as prepared as one could get in this modern day." Not only would he and his ranch hands survive a cyberattack on the electric power grid that serves Wyoming, they would barely notice it.

I asked Model to consider the possibility of the event itself. Was there anything at all transferable from his experience and environment to urban dwellers living on the nineteenth floor of an apartment building? How would they get food, water? How would they handle hygiene? Model seemed genuinely stumped. "No answers. It scares me to death just to think about it. I don't know how one would cope with that, other than that there needs to be local, state, county programs sponsored by the federal government. These are real issues, these are real concerns. It is overwhelming just to think about it." Bob Model seems like a genuinely caring man and he wants to be helpful. In a testament to his deep social values, Model stresses not survival skills nor land investment but community involvement. He cites the self-reliance teachings of the Boy Scouts, as well as organizations like Wyoming's National Outdoor Leadership School and Outward Bound.

From the living room of Bob Model's ranch these are all reasonable suggestions, but few, if any, would have an application in Manhattan or Chicago or Los Angeles. This is a region with a very small population. Emergencies are routinely addressed by the sheriff's department, augmented by volunteer organizations. The fire department is all volunteer, as are the search-and-rescue forces. In more than one sense, Wyoming is something of a time capsule, reminiscent of, say, the 1950s, when magazines such as *Look* and *Life* and *Collier's, Reader's Digest,* and the *Saturday Evening Post* celebrated an America of values and tradition that existed, in part at least, only in our imaginations. However romanticized this vision might have been, the fact is that Wyoming does have an unusually strong culture of both self-reliance and civic cooperation.

In 1995 a Harvard political science professor, Robert D. Putnam, wrote a journal article that he later expanded into a bestselling book. The theme of *Bowling Alone* was that the civic organizations that once served as such an essential part of our societal fabric were losing membership. Americans, Putnam argued, were going it alone, less inclined to participate in the civic groups that once kept communities close. That is certainly not as true in Wyoming. The website for the *Wyoming Tribune Eagle* listed meetings of clubs and organizations for the week of March 16, 2015, including such familiar names as the Lions (sunrise meeting at 6:30 a.m., noon Lions at, well, noon) and the Rotary Club, but also the High Plains Toastmasters, the High Noon Toastmasters, the Roadmasters Toastmasters, the Audubon Society, the Wyoming Mounted Search and Rescue Association, the Elks Lodge 660, the Masonic Order of the Eastern Star, and the Kiwanis Club. The list for that week alone ran to nine pages. Neighborliness is more than a slogan here; it is, as it has always been, an essential element of self-preservation in a challenging environment.

The walls inside Bob Model's ranch house bear testimony to the integral role that guns and hunting play in his life. The head of a ten-point buck hangs above a rack carrying a pair of lever-action rifles. The entry hall is a classic American montage: several pairs of western boots next to a wooden rack carrying a well-worn saddle, a lariat hanging from the saddle horn, and a couple of Smokey the Bear–style hats. A splendid American flag is draped on one wall. A couple of oil paintings of western scenes hang over a gun case tucked in a corner.

Isolated as it is, Mooncrest Ranch is not merely self-sufficient; it could, I suggested to Model, become a magnet to those in desperate need during a time of regional crisis. What about the limits of neighborliness? What would happen, I asked, if those with fewer resources came calling? "That's a great question, and a very complicated question," he demurred. "Traditional western hospitality. If they were not gangsters or thugs, they would be welcome here." It may be, given some planning and organization, that states such as Wyoming, with a very low population base, would provide sanctuary to some tens of thousands of urban refugees. Absent any such plan—and there is no evidence that it exists—the search for food and shelter is likely to become more chaotic.

I pressed the point, giving Model a worst-case scenario. "Let's say there's a band of twenty or thirty armed people who have decided that this is a good headquarters, that this would be a good place to have."

"You know," Model allowed, "we've thought about that, and I mean, we're armed here, too. We know how to take care of ourselves. So it's very difficult for me to speculate about what would happen. But for somebody coming and taking over the ranch, they're going to have a fight on their hands. My cowboys and people who work here are pretty self-reliant."

Firearms, Model continued, are "a function of where we live." He explained the day-to-day uses of his guns, which, in addition to hunting and protection from "large predators," could occasionally be needed to put down an injured or ailing horse. "So we understand the very sensitive issue of firearms. We feel that we're very responsible and we feel that they're part of our way of life."

Guns are indeed part of Wyoming's way of life. While it has the smallest population of any state in the nation, Wyoming's rate of gun ownership is, at 62.8 percent, the highest in the country. There is general agreement on those two statistics. From that point on, however, views differ. Whereas gun control advocates routinely list Wyoming as having one of the highest rates of gun-related homicides in the country (calculated on the basis of shooting murders per 100,000 of population), statistics from the Centers for Disease Control (CDC) tell a different story. For the five years 2009–2013, the CDC reported a total of 42 gun-related homicides in Wyoming, for a ratio of 1.47 gun murders per 100,000 population—the fourteenth-*lowest* in the nation. California, by comparison, had more than double that rate, while the District of Columbia, with the worst gun-related homicide rate, registered more than eight times as many as Wyoming. The District of Columbia has some of the strictest gun laws in the country, Wyoming some of the least restrictive. As with expert testimony, statistics are often selected and cited to suit preexisting opinions.

At the Park County Public Library back in town, Alan Simpson had invited his old friend Stanley Wolz to join us in the cafeteria for a bite of lunch. Like Model, Wolz rejected any trendy conception of "bugging out."

"I am not a doomsday prepper," Wolz told me. "I am an emergency-preparedness guy. I have done my planning in layers. The first part would be maybe three weeks where we wouldn't have to go out of the house. Anything after that, probably, we could probably go six months. But that's all predicated on water. If you can't get water, you can't clean yourself, you can't cook, you can't do anything. My whole thing," he continued, "is not to have a school bus that I load with five thousand pounds of stuff and go build a hut in the forest. I think that's silly."

Stan Wolz is an amiable man and, as a Mormon, is committed to sharing within his community and with those in need. He is also well armed. He has a hunting rifle with a scope, what he calls a "home defense" shotgun, and a variety of pistols. Wolz is ambivalent about when he would use one of those self-defense weapons. He is not sure whether he would shoot someone over his "bucket of beans in the basement," as he puts it. "I just think I'd rather have a firearm and not need it than not have one and need it." But he is, if anything, more realistic than Model about the possibility of the situation escalating: "I think that within six months, if things weren't straightened out adequately to have something besides anarchy, somebody's going to show up on your doorstep and kill you for your food anyway."

Martin Knapp, the homeland security coordinator for Park County, Wyoming, believes that other parts of the country are different. In urban and suburban areas he would be concerned that people might use their guns "to take from their neighbors. You know, 'Hey, we're out of food. We've gotta get more stuff.' " In Cody, Knapp described a culture in which kids go to hunter-safety class, but also in which parents educate their children in gun safety from an early age. "That's why I'm not as concerned that these people have guns."

What about outsiders fleeing the cities? I asked Knapp. How would that change the dynamic? It didn't seem to worry him: "As far as people coming in here from a couple hundred miles away, they're going to have to get enough fuel to get here. That's one thing. But somebody comes in here and pulls a pistol on somebody, 'Gimme your food' "—the prospect was mildly amusing to Knapp—"four people in the house pull out rifles and everything else, saying, 'I don't think so.' It's like bringing a knife to a gunfight. People around here, because there are so many guns, they've got ways to kind of protect themselves. And a lot of them would, without thinking twice about it."

A place like Cody might very well be brimming with "western hospitality," but this is a place that has never been put to the test of a large-scale influx of domestic refugees. How over-burdened civic organizations would respond, to say nothing of individual citizens, is unknowable. When people in Cody consider the possibility of violent criminals, they tend to put it in the context of outsiders.

Jeff Livingstone is very much part of the fabric of the city of Cody; one of a dwindling number of ditch riders, he works to deliver irrigation water to farmers. "It's one of the craziest jobs that you could ever stumble into," Jeff explained. "One of the things Buffalo Bill did for Cody was he helped promote and design a Shoshone irrigation project back in 1903 or 1904."

In the old days, ditch riders were on horseback and rode along the dry beds of the ditches alongside the farmland. These days, Jeff drives his truck to the head gates and works the valves. "I go out every morning for about two, two and a half hours." Folks tell him what they need: "They want three feet of water, three cubic feet per second." The water comes from the Buffalo Bill Reservoir and flows through those same ditches. Farmers pay a flat fee for their allotment. If they use more, they pay a surcharge.

Jeff and his partner, Pat Altringer, who is a retired nurse, aren't fond of the term "prepper," preferring "self-reliance," though they say the former is "fine" in a pinch. Self-reliance is probably a more accurate term in their case, anyway. It is simply how people of limited means in Cody, Wyoming, prepare for shortages in hard times. Jeff and Pat have a 500-gallon gasoline tank and a small generator. The generator can run the electric grinder that Jeff uses to make flour. Anticipating a variety of needs for the generator, Jeff is concerned about devoting too much of its use to grinding wheat berries. He's thinking about buying a wheat grinder that he could hook up to the old bicycle outside. He's laid in about 150 pounds of wheat berries. He also has about 130 pounds of pinto beans and 30 to 40 pounds of black beans. He and Pat learned how to can the produce they grow; whatever they don't eat immediately, they'll can.

Jeff used to be a big-game hunter and he's pretty sure he could bring in some meat if needed. "And then we've got those two nags out there," Pat says, though she may be joking. "You know, if they can't provide transportation anymore ..." The horses under discussion are grazing contentedly in a field outside the house.

Jeff and Pat are not counting on any government assistance in time of need. "And it isn't because I hate the federal government," Jeff assured me. "I think it's grown too large. But if you had a crisis like that, a grid down, that would probably be pretty immediate and pretty catastrophic. You know, they've got millions of people to take care of and major population centers that will break down a lot more quickly than rural areas like this. So I would be surprised if we ever saw FEMA in Cody, Wyoming."

Jeff is clearly proud of his connection to Cody's frontier history. In trying to explain why Cody still works as a community, he noted that Cody is 96.9 percent white and 75 percent registered Republican. He was quick to reassure me that he is not a racist. "It's just when you get that kind of ... ," Jeff began, then paused, searching for the right word.

"Non-diversity?" Pat offered.

"Non-diversity," Jeff continued, "makes for, you know, very little stress."

If you're looking for a way to sum up Cody's demographics,

non-diversity is almost perfect. There are Shoshone and Arap-
aho reservations about 150 miles south. There's a Crow reservation, but that's across the border in Montana. There are some
Mexican Americans in Cody, but the total Hispanic population
is 2.22 percent. I looked up the numbers after talking to Jeff,
but he was pretty much on the money. African Americans, he
conceded, are essentially missing from the mix. "I think like
0.78 percent. And the ones who come here, from my observation,
are welcome. It's just hard to be a black in a 96 percent white
community. I knew Black Jerry, [who] was a black cowboy that
came out here and cowboyed for my cousin Lee Dan, who's an
outfitter out here. He's great. Got along fine. It's just culturally
it's so different." Jeff had, in fact, inadvertently exaggerated the
number of African Americans in Cody. The U.S. Census Bureau puts the percentage at an even more modest 0.2 percent. If
it has occurred to Jeff that Cody's "non-diversity" may be the
natural outcome of a social environment that does not welcome
African Americans (Black Jerry notwithstanding), he does not
mention it.

This is how Model's claims of "western hospitality" and
Knapp's boast that "people are so good about working together
here" can coexist with Wolz's prediction of anarchy within six
months.

The disconnect between New York and Cody, or Cody and
Los Angeles, is even greater than it may seem. Underlying all
expectations of survivability in a major city like New York is the
assumption that underpopulated places such as West Virginia
or Wyoming could, in extreme circumstances, absorb a couple
of hundred thousand urban refugees. And perhaps they would,
because the residents of those areas really do take notions of

neighborliness and community values seriously. But when Joe Nimmich of FEMA and former DHS secretary Michael Chertoff speak blithely of evacuating several million people from a city like New York, there is really no concept of where they might resettle all these refugees. Insofar as the residents of a town such as Cody, Wyoming, have maintained their traditional values, they have done so in an environment of what Jeff Livingstone described with searing if unintentional candor as "non-diversity." Perhaps if an urban exodus was part of a carefully imagined and well-thought-out plan, it might have a chance of success. If such a plan even exists, it might serve the greater good to let it be seen before it is needed. To just assume, however, that the underpopulated rural regions of the United States are inclined or even able to absorb tens or hundreds of thousands of urban refugees—white, black, brown, many of them poor—is to place too much reliance on the notion of neighborliness.

Which brings us to the Mormons. While conceding all glory to the Almighty, they are firm believers in the precept that God helps those who help themselves, and no group in the country approaches anything like the extraordinary scale and geographic scope of their efforts. Next we'll explore the singular degree to which they have prepared for disaster—and its relevance for the rest of us.

16

....

The Mormons

In the sweat of thy face shalt thou eat bread, till
thou return unto the ground.
—GENESIS 3:19

What intrigued me about the Mormons was that their plan en-
gages an entire community. When I first heard about it, it didn't
go much beyond the notion that Mormon families were encouraged to keep a three-month supply of food and water on hand.
There were also faintly ominous suggestions that the Church
of Jesus Christ of Latter-day Saints had gigantic, strategically
located storehouses filled with supplies. I envisioned armed
guards, electrified fences, and barbed-wire enclosures.

   I contacted the church's public relations department in Salt
Lake City and asked for the appropriate contact information.
A little church history, I thought, and then a quick visit to a
local warehouse. There is, after all, an enormous and imposing Mormon Temple less than half an hour from my home in

Maryland. The message came back: the church would be happy to help, but I would have to fly out to Utah. I phoned Senator Orrin Hatch and explained the problem. I thought the senator, a prominent Mormon and someone I've known for many years, might be able to put in a good word and save me the trip to Salt Lake City. Hatch called me back a couple of days later to let me know that I would be receiving a call from Henry Eyring, a church leader in line for the presidency of the LDS church. Senator Hatch himself seemed slightly awed. "This is one of the most important men in the world," he told me. Hatch himself is president pro tempore of the U.S. Senate, a title bestowed on the longest-serving senator of the majority party—not someone I would have thought was easily impressed.

Henry Eyring, when he called, couldn't have been more gracious or reassuring about the cooperation I would receive in researching this book. I would have to fly out to Salt Lake City, though, and I should probably allow two or three days for the visit. If I was going to investigate the Mormon approach to disaster preparation, it appeared that it would take more than a quick drive around the Washington Beltway and an afternoon's worth of research.

The LDS church is not a casual operation. The office headquarters in Salt Lake City are linear and somewhat stark, like something out of an Ayn Rand novel. The interior space, where the offices of the church leadership are located, bears more than a passing similarity to the antechambers guarding the offices of the secretary of defense or the chairman of the Joint Chiefs of Staff at the Pentagon. Nor is that the only echo of military structure. The church is a highly disciplined, hierarchical organization. There's no uniform, but, as a matter of courtesy and mutual respect, the men customarily come to work dressed in dark suits, white shirts, and ties. And it is, if anything, even more tradition-bound; unlike the military, which has elevated at least some women into its highest ranks, those who govern the LDS church—the members of the First Presidency, the Quorum of the Twelve Apostles, the Presidency of the Seventy, the First Quorum of the Seventy, the Second Quorum of the Seventy, the Presiding Bishopric, slightly more than a hundred in total—are all men.

If all of this appears incidental, it is not. Like many devout believers of other faiths who have learned over generations not to take tolerance for granted, Mormons have been conditioned by prejudice. They are encouraged to prepare for the days of tribulation as a matter of religious doctrine, but also as a direct consequence of historical experience.

Though the LDS church is still the object of a certain degree of bigotry, these days bemused curiosity has largely replaced outright hostility. Where most of our mainstream religions can now cushion the supernatural aspects of the miracles and mysteries of their faiths behind an accumulation of millennia, the Mormon belief system is not yet two hundred years old. Joseph Smith's translation of the Book of Mormon, written on plates of gold and revealed to him by the angel Moroni in upstate New York during the early 1820s, served almost simultaneously as the foundation of a struggling new religion and the object of ridicule, prejudice, and brutal persecution. When, during many of the religion's early years, that hostility manifested itself in the form of physical violence against church members, preparing for disaster became an essential element of survival; this ultimately matured into a matter of doctrine itself. To gain a clearer understanding of how those early experiences shaped the modern church's approach to disaster preparation, I met with Richard E. Turley, one of the official historians of the LDS church.

After the church's legal establishment in 1830, Turley explained, the small body of two or three hundred believers resettled twice in short order. Responding to a divine revelation, Joseph Smith first moved his followers from upstate New York to Ohio, where the Mormon community barely began to flourish before Smith announced the establishment of a parallel headquarters for the church in Missouri, in a new city to be called Zion. There was almost immediate friction in Missouri. The Latter-day Saints claimed to talk with God, which didn't go over well with local Protestants in the 1830s. In addition, Missouri was a slaveholding state. Most Latter-day Saints, coming from the Northeast, opposed slavery, welcomed blacks into the church, and during those early years even ordained black men to their priesthood. In 1833, vigilantes drove the Latter-day Saints out of Jackson County, Missouri, across the Missouri River into Clay County. There followed a few relatively peaceful years, and then the Missouri legislature moved to isolate the Mormons by creating a separate county for them. The intent was that this new county would serve as a ghetto, the equivalent of an Indian reservation. When Mormons began to stray from the reservation, vigilante attacks resumed. These escalated, explained Turley, until "there was a civil war, what's called the Mormon-Missouri War of 1838, with Latter-day Saints saying, 'You know, we've been pushed and pushed and pushed. We're Americans. We have rights, and if we don't put our foot down, this is just going to keep going on.' So they decided to fight back, and it resulted in a massacre of a Mormon village."

This is history through the church's prism, but that's exactly what I was looking for. From Joseph Smith's point of view, what he and his followers were facing was another existential threat to the Mormon community. Following another skirmish between the Mormons and a state militia group, the governor of Missouri ordered the "extermination" of Latter-day Saints from Missouri. In the winter of 1838–39 the Mormons crossed the Mississippi River into Illinois.

There the church flourished—briefly. With the establishment of the city of Nauvoo, Joseph Smith created a formidable power base both for the church and for himself. Smith continued as leader of the LDS church but also became the mayor of Nauvoo, the head of its municipal court, and, perhaps most significant, lieutenant general of the five-thousand-man Nauvoo Legion of the Illinois state militia. His city was, at the time, the size of Chicago, each with a population of about fifteen thousand. Joseph Smith had accumulated real power—and powerful enemies outside the church who feared his growing influence in Illinois. There were even published reports, one of them in the newspaper of a dissident Mormon group, that Smith had reestablished plural marriage, a practice among certain Old Testament kings and prophets. Smith loyalists destroyed the dissident newspaper's press. But for Smith's enemies outside the church, it had all come together: the rumors of polygamy, the "assault on a free press," and most of all, concern over Smith's growing power. It resulted in his arrest in 1844 on a charge of treason and conspiracy. Joseph Smith and his brother and designated successor, Hyrum, were brought to Carthage, Illinois, where a lynch mob stormed the jail and fatally shot both men.

Leadership of the LDS church then passed into the hands of Brigham Young, and it is here that the history and culture of the church were set on a different course. When attacks against Mormons of Nauvoo resumed, Young adopted a pacifist policy. He declared, as the historian Richard Turley paraphrased, "Let them pull the trigger, not us. We're just going to keep retreating." Young led the Mormons across the Mississippi River and headed west. Images of Mormon families pulling handcarts loaded with all their belongings are iconic in church history. It was 1847 when Brigham Young finally settled his followers in the Salt Lake Valley. "He wanted someplace where they wouldn't be persecuted," explained Turley. "A place that no other Europeans wanted." It is where the heart of the Mormon Church beats to this day.

Most western towns spread out in random fashion. Salt Lake City was laid out in carefully planned ten-acre blocks, divided into one-and-a-quarter-acre lots to which settlers were assigned. Streams were diverted from the surrounding mountains so that each street had a supply of fresh water. There wasn't any hardwood, so people built with adobe. By the late 1850s travelers describe coming out of the canyons and seeing beautiful homes, neatly arranged with gardens and trees. That physical orderliness would be an early indicator of the church's own structure.

It may seem curious that in gathering information about the impact of a cyberattack on one of the United States' electric power grids, I have spent the better part of a chapter sketching out the cross-country journey of the earliest Mormons. But it is a necessary prerequisite for understanding why the church is so focused on preparing for the unexpected. What Brigham Young told his followers in Salt Lake City created a mindset that informs Mormon behavior to this day: "If you are without bread, how much wisdom can you boast, and of what real utility are your talents, if you cannot procure for yourselves and save against a day of scarcity those substances designed to sustain your natural lives?"

One can easily become entangled in a theological discussion of the Latter-day Saints and their approach to the "end of days." That may or may not be a motivation for the church's ongoing emphasis on preparation. In any case, what they have achieved is extraordinary. No group of comparable size comes close to matching the scale and organizational discipline of the Mormons' efforts to prepare for whatever catastrophe may come. Their example is hardly an easy one to follow, but it serves as a model of what can be done.

State of Deseret

If you are prepared you shall not fear.
—BISHOP GÉRALD CAUSSÉ

Even the most elaborate framework requires a solid foundation. For the Mormons, this means starting with families, who are encouraged to prepare, over time, for unspecified emergencies. They are urged to gradually set aside enough food, water, clothing, and money to sustain themselves for three to twelve months. Ezra Taft Benson, who served the Eisenhower administration as secretary of agriculture before becoming the church's thirteenth president, framed the issue in thoroughly pragmatic terms. "Have you ever paused to realize what would happen to your community or nation," he asked in 1980, "if transportation were paralyzed or if we had a war or depression? How would you and your neighbors obtain food? How long would the corner grocery store—or supermarket—sustain the needs of the community?" He recommended storing at least a year's worth of supplies.

My visit to Salt Lake City was meticulously choreographed, and it was suggested that I share a home-cooked meal with a local family. If Norman Rockwell had chosen to paint the all-American Mormon family, he could very easily have ended up with the Turleys. Kate and Trey Turley are wholesomely good-looking, and there is an impish quality about their three boys. Their pantry, which Kate estimated contains enough stored food to feed the family for about six months, looks well organized but less overwhelming than I had imagined a six-month supply would be. (Big water containers!) That's where Kate picked out the ingredients for our dinner that evening: chicken with pasta, frozen bread crisped in the oven, salad, and nonalcoholic apple cider. (The Turleys are, after all, a Mormon family, so there was no alcohol, tea, coffee, or caffeinated soft drinks.)

We sat down after dinner to talk about a few of the aspects of their lives that identify them as Mormons. Most of what they described would not seem unfamiliar in a conservative Christian, Jewish, or Muslim home. Self-discipline and charity are high on the agenda. In their case, that entails giving 10 percent of their income to the church. (Tithing is mandatory for church members in good standing, but whether the 10 percent applies to gross or net income is left up to the individual. The Turleys opt for the more generous option of tithing from their gross income.) Kate explained that the boys are expected to tithe from whatever they earn: "So if they mow the lawn, they get $20 for the lawn, they pay $2 to their tithing." On the first Sunday of each month, there is an additional fast offering. "Depending on where you sit financially," said Trey, "the idea is to try to provide the funds that you would have spent on those meals that you fasted. But as you do better, there's certainly nothing against adding funds to that."

"We try to give as much as we can on that day," added Kate. "So we go to bed Saturday night, we start our fast after dinner with a prayer, and then we skip breakfast and lunch and then we eat dinner Sunday night." Since each of the boys is now over the age of eight, they all fast.

On Monday evenings they meet as a family to discuss important issues. It's not a mandatory church practice, but it's strongly recommended. These conversations can focus on a fire drill, charity, injunctions against smoking and drinking, or disaster preparation. Again, elements of this might be familiar to devout members of other faiths. Like many observant religious groups, practicing Mormons follow top-down instruction to maintain a set social structure, and almost all religions stress charitable giving. What distinguishes the Mormons is their extraordinary focus on the integration of self-sufficiency and charity. That carefully layered structure is what gives the LDS church its impact and efficiency.

Mormon families belong to a ward, as the church calls its congregations. There is no hard-and-fast membership number, but in a high-density Mormon population, a ward can be as large as four or five hundred people. Wards are presided over by a bishop and two counselors, in a structural pattern (units led by a presiding body of three) repeated throughout the church. Above the ward level is the stake, analogous to a diocese in the Catholic church. The number of wards in a stake can vary but on average is about ten, so there might be as many as four or five thousand people in a stake, with a president and two counselors presiding over each stake. The triumvirate at the pinnacle of the ecclesiastical hierarchy is known as the First Presidency, comprising the Prophet or President, a First Counselor, and a Second Counselor. Just beneath them is the Quorum of the Twelve Apostles, then the First Quorum of the Seventy, and then the Second Quorum of the Seventy.

This organizational pattern is repeated, literally, around the world. However unfamiliar its governing structure may be to nonmembers, the Church of Jesus Christ of Latter-day Saints is a vast, efficiently run enterprise, claiming a worldwide membership of more than fifteen million, of whom roughly six million are in the United States. Describing its intricate hierarchy is, in the context of this book, useful only in this sense: one would be hard pressed to find a large religious organization with more precise systems of communication and oversight.

Each ward is expected to have its own emergency plan, designed to deal with the sorts of natural disasters peculiar to the region, and a local bishop's placement within a military-style chain of command gives him and those in his care access to an incredible network of resources. In case the electricity is out and the normal phone system is no longer functional, satellite phones have been pre-positioned with stake presidents and a network of ham radio operators has been set up. If cars aren't working or if gasoline is scarce, local bishops have plans to send messengers by bicycle or foot. A ward bishop knows who in his community is vulnerable and where he can find those with the particular skill sets to assist—those who can, for example, provide medical assistance, or help with rudimentary repairs. According to Steven Peterson, managing director of the church's welfare services, the bishop's recognized authority within the ward enables him to organize the community at the local level. Peterson expressed complete confidence in this "chain of authority," declaring that "in the situation where communications are largely eliminated and it's not possible to connect with other leaders, we firmly believe that the individual members who have storage and food and preparation in their homes, led by local bishops and stake presidents, would immediately join together and figure out how they could best care for each other."

The finely calibrated, professional operation this national network represents can, and has on occasion, outstripped Washington's own disaster response machine. In a 2007 article for *Mother Jones*, Stephanie Mencimer recounted the LDS church's reaction to Hurricane Katrina in 2005. It was, she wrote, "a performance that put the federal government to shame." The New Orleans branch of the LDS church had evacuated all but seven of its approximately 2,500 congregants before the storm even hit, largely because the church had created an automated telephone emergency warning system that alerted all its members, instructing them to get out of town and telling them where to go. "While FEMA was floundering," Mencimer wrote, "the church dispatched ten trucks full of tents, sleeping bags, tarps to cover wrecked roofs, bottled water, and 5-gallon drums of gas from its warehouses to New Orleans and other hard-hit areas. The supplies were distributed in an orderly fashion to people who desperately needed them."

There is an entire branch of the church hierarchy dedicated to dealing with the administration of the church's physical assets—its 144 temples and 28,000 buildings worldwide, as well as the torrent of resources it produces, sells, and distributes. These temporal affairs are the responsibility of the Presiding

Bishopric, of which Gérald Caussé, a Frenchman whose parents were Mormon converts, is First Counselor. "There is scripture that we refer to in the church," Caussé told me, "that says that if you are prepared you shall not fear. This is at the center of our religion, this scripture, to be prepared. So a lot of what we are doing is teaching people, helping people get prepared in their families for anything that comes."

"Anything that comes" doesn't have to be a disaster on the scale of Katrina, or indeed any kind of collective emergency. Where a family is no longer able to deal with its own setbacks, whether that be a lingering illness, a sudden death, or perhaps the loss of a job, it is still expected to turn to its ward bishop. It is through a bishop's "recommend" that a ward member can be granted access to what is variously referred to as a "bishop's storehouse" or "Lord's storehouse."

A recommend, which can be bestowed or withdrawn, is an incredibly powerful tool for influencing behavior. To a ward member experiencing hard times, the bishop's storehouse can be an indispensable resource. "It's like a grocery store," Caussé explained, "but you don't have to pay. You know, there is no cashier and you cannot pay at the end. You choose to come with your recommendation of the bishop and you will take whatever food you need for your family." A recommend is not an open-ended license to load up a shopping cart, but it is a generous extension of charity for the time that the bishop determines it will take a family to become self-reliant again. Those availing themselves of the bishop's storehouse are expected to volunteer their own time to work at the storehouse for a few hours each week. Several of my hosts stressed the high priority that the church places on volunteering, both as a means to strengthen the fabric of the community and as a way for those receiving help to restore their own self-esteem.

I saw this effort to treat everyone as useful and functional firsthand during my visit to a bishop's storehouse. I was struck there by the presence of a young staffer who was clearly challenged. His name tag identified him as a church elder. He was, it was explained to me, incapable of fulfilling the duties of a missionary overseas, but with supervision he could assist with simple tasks at the storehouse to fulfill his obligation. It would be the first of several instances in which I witnessed a determination not to consider or treat any member of the community as anything other than productive.

The storehouse I visited is part of a sprawling network. Think of 111 mini-marts spread around the United States and Canada. Like any supermarket chain, the individual stores need to be replenished, for which there are four large central warehouses in the United States and one in Canada. Those warehouses, in turn, are resupplied from an enormous complex in Salt Lake City. Only a chain of stores on the scale of Costco or Walmart would need a facility of this size: building after building, thirty or forty feet high, pallets stacked floor to ceiling with everything from food supplies to shelves of truck tires. Some buildings are air-conditioned, others refrigerated. There are giant generators ready to provide electricity in the event of a shutdown and underground tanks of diesel fuel with an estimated storage capacity of 250,000 gallons. The diesel also fuels the trucks that service the church's national distribution system.

Those trucks, too, are owned and operated by the church— part of LDS's proprietary trucking company, Deseret Transportation. An online ad for drivers specifies that they must hold a federal commercial license, meet all Department of Transportation state and local requirements, be able to back trailers into a dock for loading and unloading, and be a member of the Church of Jesus Christ of Latter-day Saints. A prospective trucker for Deseret Transportation must also be "currently temple worthy," which is no simple task. The concept of being "worthy" to enter the Lord's house has its roots in the Old Testament, the 24th Psalm: "Who shall ascend unto the hill of the Lord? Or who shall stand in his holy place? He that hath clean hands, and a pure heart." The Mormons take that literally and require that each church member be interviewed by priesthood leaders at least once every two years. Each member is asked the same questions, dealing with, among other issues, chastity, tithing, church attendance, honesty, keeping the covenants, and sustaining the president of the church. If a member is found to be worthy to enter the temple, he is issued what is called a "temple recommend." This is a card signed by his bishop, valid for two years and granting the holder admission to the temple.

The relevance of a temple recommend to the qualifications of a trucker is not as obscure as it may seem. Deseret Transportation, I was assured, has the best safety record of any trucking company in the country. I cannot vouch for that, but it makes sense. A Mormon truck driver who is temple worthy can reliably be assumed to be drug and alcohol free, and not even to rely on caffeine as a stimulant. Starkly put, the temple recommend ensures the church leadership a degree of top-to-bottom control over anyone who would consider him- or herself a practicing Mormon.

Deseret trucks transport product not only to all bishop's storehouses but also to more than one hundred church-run home storage centers, where church members and outsiders alike can buy prepackaged nonperishable goods. The trucks are available, the drivers are reliable, and there is a vast emergency reservoir of fuel. Even this extensive infrastructure, however, merely hints at the scope of the enormous, self-sustaining food chain that supplies the Latter-day Saints.

Don Johnson is director of production and activities of the LDS welfare program. "We have fifty-two farms, ranches, and orchards; twelve canneries and processing plants," he told me. Together these constitute a self-sustaining conveyor belt carrying enormous quantities of food into the vast distribution system that supplies the church's emergency preparedness network directly and, through sales beyond the network, indirectly. It's difficult to quantify how much food they give away through the bishop's storehouses, but Johnson estimated that it works out to about $145 million worth annually. Approximately 60 percent of what is grown and produced and processed by church subsidiaries is sold on the open market. In the event of a massive crisis, though, everything could be consolidated to provide resources for the church and its members.

The church is independent of any outside supplier. What it donates, what it sells, and what it puts into storage all come from within a single ecosystem. There were enormous silos outside the facility where we met. The harvest had just ended, so the silos were filled with about twenty-four million pounds of hard wheat, soft wheat, and durum wheat. Bill Dutton, the manager of the Deseret Mill and Pasta Plant, showed me two gigantic, gleaming pasta-making machines that had just been imported from Italy. The pasta-making operation alone—one machine for the long goods, like spaghetti, the other for the macaroni, elbow, and small goods line—represented an investment of somewhere between $18 million and $20 million. Their projection was for an average of four million pounds of pasta annually.

In Elberta, Utah, about seventy-five miles south of Salt Lake City, I was met by David Secrist, vice president of cattle operations for Ag Reserves, Inc., a privately held subsidiary of the Mormon Church. He and I were standing on a hill overlooking a long, narrow valley, and he nodded in its direction. "We have about ten thousand irrigated acres that we grow feed [on] for the cows on the farm." Dave's use of the term "cows on the farm" may conjure up an image of placid herds of Holsteins chewing their cud in rolling fields, but this is a more industrial enterprise. These cows are, most of the time, under roof in open-sided barns and supplied by feed that lies fermenting under tarps in open yards. There is a small mountain range of feed: enough forage and corn base (which will be mixed with mineral additives) to feed five thousand mature cows and four thousand young heifers for a year. Each young heifer has its own plastic "kennel" to protect it from the elements. The lines of kennels have the look of a neat, if sprawling, military camp.

The lactating cows are milked three times a day, for a total of about 84 pounds of milk per cow per day. The five thousand cows on this facility produce 420,000 pounds of milk a day. That's about 50,000 gallons of milk per day. It all goes into the production of fluid milk, cream, ice cream, sour cream, butter, and cheese.

Cows usually "leave this operation" after they've had three to five calves and are six or seven years old, at which point they become a part of the meat-generating part of the food chain.

The torrent of dairy and beef products flows into the same production stream that carries millions of pounds of wheat, miles of pasta, tons of fruit from the orchards, and thousands of gallons of honey from the apiaries.

Many of the church's companies and operations carry the name Deseret. It's a term used in the Book of Mormon, meaning "honeybee," and it has symbolic resonance within the church. Honeybees, too, work and live within a self-sufficient, collaborative, and highly productive community. Mormon settlers at one point proposed calling the territory around Salt Lake City the State of Deseret.

With their vast resources, ongoing production, and wide distribution network, the Latter-day Saints are prepared for just about anything. They have responded to postwar shortages, earthquakes, hurricanes, and tsunamis. But there has never been anything quite like the loss of an electric power grid over an extended period of time. Caussé conceded as much, saying that, in the event of a disaster of truly national scale, "I hope we have enough to take care of the members of the church. We hope that all our families have enough to care for themselves and to help their neighbors. But there is no way we can take care of the whole country."

What would they do if, in a time of widespread national shortages, others tried to take it all away from them?

Constructive Ambiguity

I cannot imagine turning away hungry people
when we have food at our house to share.
—ELIZABETH TAGGERT, LDS MEMBER

If they came up on my porch, armed . . . I'd
probably shoot them.
—STAN WOLZ, LDS MEMBER

In the event of a national catastrophe, if an electric grid went down, every bishop's storehouse—not to mention their supply chain—would seem an inevitable target for looters. It's an uncomfortable topic, painful to contemplate; but with their history of having been driven from pillar to post and their disciplined culture of preparing for the worst, it is hard to believe that church leaders in Salt Lake City haven't considered the issue.

As president of the Presiding Bishopric, Bishop Gary E. Stevenson sits very near the top of the church hierarchy. The temporal affairs of the entire church are his direct responsibility.

"What if," I asked Stevenson, "the church institution, in terms of its warehouses, becomes a target?"

"We rely on who we rely on every day in a scenario like that. I think we have to rely on the reaction of those who have the responsibility to police the citizens, and that doesn't fall under us as a church. We've never built a preparation model that would talk about arming ourselves, or arming members of the church to defend . . ."

I assured him that I was not suggesting the church had any plans to militarize.

"I know you're not," he said, "but I'm suggesting that we don't."

The issue of guns and self-defense is a minefield for any religious organization. It is particularly so for the LDS church, within which there is a historical sensitivity to the subject of armed violence. During the years of their trek across the country, Mormons were both victims and perpetrators, albeit more often the former than the latter. Active self-defense remained an issue into the second half of the nineteenth century, even after the church put down its roots in Salt Lake City. The open acknowledgment of plural marriages among Mormons led to such tensions with other settlers in the Utah territory that in 1857 President James Buchanan sent the army to maintain order. With memories of what had happened to Mormons in Illinois and Missouri still fresh in their minds, LDS members prepared for an invasion. What ensued was somewhat hyperbolically called the Utah War. It hardly rose to the level of a war, but it was marked by a particularly horrible act of violence, when a Mormon militia unit slaughtered a wagon train of men, women, and children on their way to California, in what came to be known as the Mountain Meadows massacre. It has taken the church a long time to come to terms with the aftermath of that incident, and it's only in recent years that it has acknowledged the Mormon militia unit's complicity.

Even the most benign neighborhood security patrols would run the risk of exhuming such images among church critics. The Church of Jesus Christ of Latter-day Saints has had a difficult enough time being accepted into the mainstream of American life without reviving volatile images of a highly organized, self-sufficient, and heavily armed entity. It is totally understandable that the church leadership won't even discuss the option of defending its resources with an armed force. Nothing would more completely isolate the church. Still, certain internal contradictions remain unavoidable.

The Pew Research Center estimated that there are between 270 million and 310 million guns in the United States, and in western states with a large Mormon population, including Utah, Idaho, Nevada, and Arizona, guns are an accepted and normal part of many, if not most, households. The LDS church has set world-class standards for disaster preparedness and imbued its membership with a sense of individual and group responsibility. Its strategists have established a communications network that would survive anything short of a nuclear attack. It requires an almost deliberate act of obliviousness on the part of church leaders, for whom the gathering and listing of every potential resource is an essential ingredient of disaster preparation, to avoid any discussion of guns and self-defense. This is an organization that, in almost every other respect, stresses its self-sufficiency, its independence, its reluctance to depend on government assistance. The issue of self-defense, though, is toxic.

"People protect their assets," said Stevenson, "and the way that we rely on protecting our assets as a society, in our country, the adequate way of protecting assets has just been through the state, through the police. And as a church, as an institution, I think, we have always felt comfortable that the state is going to provide the kind of protection that we would need."

Blaine Maxfield is the church's chief information officer. After giving me an exhaustive rundown of the church's emergency communications planning (which runs the gamut from Internet and social media and texting to those carefully distributed satellite phones and ham radios), Maxfield reaffirmed that no contingency exists for the church's own defense. "None of our plans contemplate, from our perspective, us defending ourselves. We're relying on all government agencies, really, to help protect our members of the church." Maxfield defended this position wholeheartedly, adding, "We're suggesting that each individual member, they rely on their relationship with their Father in Heaven and know exactly what best to do. And I know that might sound trite to you, but it isn't. From our perspective, we believe that they'll know exactly what to do to help with their families."

This is a church that provides precise guidance on almost all matters declining to take a position on what is clearly among the most controversial of issues. This is a church that all but demands self-sufficiency from its members urging a passive reliance on law enforcement agencies and government. Is it possible to square that circle, to reconcile the seemingly irreconcilable? Well, sort of. As Maxfield suggested, the answers came from individual members. Christopher Jameson Taggert and Elizabeth Stoddard Taggert live in Cody. They met at Brigham

Young University on their first day at school. "My mom told me I shouldn't date him," Elizabeth told me, "because I would end up living in a small town in Wyoming, which is exactly what happened." They have been living in Cody—"very happily," said Elizabeth—for the last twenty-seven years. Christopher's great-grandfather was among a number of Mormon settlers lured out to Cody by Buffalo Bill more than a hundred years ago with a promise of free water rights. There is now a thriving LDS community in Cody, and disaster preparation, Elizabeth recalled, has been a lively topic of discussion in their ward.

"We just had a list that went around the church just within the last month, and they were asking us to list our resources. They wanted to know, basically, just who has a flatbed, who has generators, who has, maybe, some gasoline stored, who has a well, what kind of talents they have that we could—who can sew, who can, you know, help in case of an emergency." What doesn't seem to come up in these church-initiated conversations is the subject of protecting resources.

"So what happens," I asked, "if some people show up at your front door, and some of them may even be carrying weapons, and they say 'give us what you've got'? What's your response?"

Chris answered, " 'Come take it.' You know, in our faith, I don't think this is a physical commandment. I think it's a spiritual commandment. I think the Lord's saying, 'Do this so that if something happens you will be prepared, but I think the real test is how you treat the stranger that comes knocking at your door.' "

They didn't budge, even when I pressed the issue. "What if the number grows, and there are thousands of starving people, and the word gets out: 'Chris and Elizabeth are really good people. All you have to do is knock on the door. They'll give you a week's supply of food'?"

"I can imagine myself in the situation," said Elizabeth, "and I can imagine weeping as I was giving my wheat away to people, knowing that it might mean that we would not be able to survive as long. But I also cannot imagine turning away hungry people when we have food at our house to share."

"If you think that this life is all there is," added Christopher, "it's catastrophic. We believe that there will be a resurrection, that you will live again. So, in a sense, the worst thing in life is not death."

It's not easy maintaining a cynical outlook in the face of what appears to be a thoroughly charitable mindset, but a lifetime as a journalist is something of a defense mechanism. The Taggerts do not have a houseful of guns. Many Mormons do, and they, at least, are not under the impression that their church leadership back in Salt Lake City has issued any directive preventing the use of those weapons.

Henry Kissinger liked to call it "constructive ambiguity." As a diplomatic device, it provides the kind of language that permits both sides to an agreement to attach subtle but significant differences of interpretation to language that has been left deliberately vague. In this instance, church leaders have repeatedly and emphatically refused to spell out any defensive measures or preparations—yet neither have they issued any prohibitions against them. This is not a matter of oversight. The Church of Jesus Christ of Latter-day Saints will never be confused with a congregation of Unitarians; the Saints are quite explicit about what is permissible behavior and what is not. In every other aspect of disaster preparation, LDS leaders stress cooperation with the state and collaboration with its agencies, but never surrendering the initiative. The church runs its own affairs. If it wanted to rule out the use of guns for any purpose other than hunting, it would do so. Instead, in Blaine Maxfield's words, the church "is suggesting that each individual member rely on their relationship with their Father in heaven and know exactly what best to do."

"Did they show you the manual that the church puts out?" Stanley Wolz, also a Mormon living in Cody, Wyoming, raised the question in the context of our conversation about self-defense in a time of crisis. "It's about that thick," Stan said, holding his thumb and forefinger about an inch apart, "and it covers everything—shelf life, how to store, how to can it, whatever to do. This last version of it was the first time I've seen mention of a firearm in there, and it didn't say what to have a firearm for. It just said you should think about having a firearm. So I think, personally, that their direction probably was for food gathering, if you were in an area where you could actually hunt. But . . ."

"But it's not clear what it's for," I said. "It's ambiguous."

"It was there," said Stan.

Constructive ambiguity.

....

Solutions

For every complex problem there is an answer that
is clear, simple and wrong.
—H. L. MENCKEN

Not even the Mormons have focused particular attention on preparing for the aftermath of a disabled grid. Their emphasis on disaster preparation adapts to local and regional realities but remains agnostic as to the precise nature of an impending catastrophe. In this respect their approach mirrors that of the Department of Homeland Security, which also generalizes its recommendations. That, however, is where the similarity ends.

Homeland Security proposes that families settle on a predetermined meeting place and that they equip themselves with sufficient food, water, appropriate clothing, money, and medicine to survive seventy-two hours—and yes, of course, the radio, a flashlight, and adequate batteries for both. It is not nearly enough. It is based on outdated assumptions that are barely adequate in the wake of natural disasters. The loss of electricity for tens of millions of people, extending over many weeks, requires something altogether different. The greater the level of self-sufficiency and the larger the number of social networks able to function independently for at least a week or two, the more successful government relief efforts will ultimately be.

The LDS church has established a model that makes good common sense, one that serves to support families in times of illness or unemployment, natural disaster or international crisis. It is designed to cushion families during hard times over an extended period. Certainly most families cannot afford to immediately lay in a six-month supply of food and water. Too many families lack the resources to meet even their daily needs. But if those who can afford it take on the responsibility of longer-term survival, supplies available to emergency management agencies can be reserved for the very neediest. Many urban dwellers, living in small urban apartments, lack space, but when what's at stake is survival, it's astonishing how much can be tucked away in small spaces. To establish a foundation with long-lasting, nourishing foods that have sustained needy families for generations—rice, wheat berries (and the grinder to make flour), beans—and large containers of water seems ridiculous in times of plenty, but it can become the difference between survival and starvation during an extended crisis. True, the wheat berries and grinder are not likely to find many converts among city dwellers, but the goal is to build up a supply of nonperishable goods, small amounts at a time. These are measures to be undertaken gradually, over time. Eventually the supplies become part of a natural pattern—rotation of the older food into a pattern of daily consumption, always to be replaced with fresh supplies.

What will, for most people, be the most difficult to replicate in the Mormon experience, however, is the intricately organized community, existing on both the local and national levels. There are well over two thousand Community Emergency Response Teams (CERTs) throughout the country. They are affiliated with FEMA and provide a useful structure for implementing disaster relief, but they don't have much of a presence in America's cities. There is, for example, only one CERT in the nation's capital. Still, it is a place to start, and if you go to the CERT website, you will find the name and contact information for the organization nearest you. Many religious communities already have a structure and sense of connection in place, as do any number of social and civic organizations. Some of these have already made thorough preparations for disaster relief. For those that have not, but where organization and a sense of community are in place, it should be a relatively manageable matter to modify what already exists.

Many of us have lost the art of neighborliness. What comes more easily to rural communities has atrophied in urban apartment complexes. But the directions for even loose associations are relatively simple: locate and establish the needs of the most vulnerable, determine the skills and assets of those willing to share either or both. (I know, it comes dangerously close to the old Marxist dictum "From each according to his abilities, to each according to his needs." Desperate times call for desperate measures.) Once disaster strikes, it is already too late. This may be an ideal opportunity to simultaneously address a potential crisis and an existing need. At a time when many police and sheriff's departments have become alienated from the communities they serve, law enforcement officers have an opportunity to demonstrate their commitment to protecting the people they serve. Law enforcement, fire departments, and teams of emergency medical workers are the ideal agencies to draw communities together in disaster preparedness. The social catastrophes that will emerge in the wake of a successful cyberattack on a grid will not easily lend themselves to grassroots solutions. But when a population's mindset is focused on social connections and basic solutions, it lays a foundation on which government can build. As history demonstrates, and as we will see in a later chapter, preparation, even when misdirected, produces unexpected dividends.

There is, unfortunately, a whiff of defeatism about preparation. It implies the inevitability of an impending catastrophe when time, effort, and money might more proactively be expended on prevention. The one should not preclude the other; the overall utility of preparing for hard times, with a rotating larder, participation in a social network, and the establishment of a financial safety net, is eminently adaptable and useful even in the absence of catastrophe. On the other hand, defending against a cyberattack is something that only a coalition between government and industry can even attempt. So it is to be expected, perhaps, that government's primary emphasis remains on prevention.

As is so often the case when we find ourselves confronting intractable problems, committees are formed. In July 2014 SIFMA, the trade group that represents the money industry—banks plus asset management and securities firms—"proposed a government-industry cyber war council to stave off terrorist attacks." SIFMA retained former NSA director Keith Alexander to facilitate the joint effort, which would bring together industry executives and deputy-level representatives from at least eight federal agencies. As outlined in an earlier chapter, CEOs from the electric power industry already meet three times a year with senior White House officials as part of what is called the Electric Sector Coordinating Council. This outreach to the federal government is a measure of rising alarm in the ranks of industry and big business, but it is only in the past year or so that the fear of government interference appears to be losing ground to the need for government protection. We are in a period of dynamic change.

Widespread recognition of the vulnerability of our power grids already exists. Lots of smart people are already offering partial remedies and grappling with solutions. But there is not yet widespread recognition that we have entered a new age in which we are profoundly vulnerable in ways that we have never known before, and so there is neither a sense of national alarm nor the leadership to take us where we need to go. Our national leaders are in a precarious place. They recognize the scale of danger that a successful cyberattack represents. However, portraying it too graphically without having developed practical solutions runs the obvious risk of simply provoking public hysteria. For the moment, we are immersed in partial solutions.

Industry and government are creeping toward an alliance of sorts. It comes in the wake of daily cyberattacks and probes, and of having already spent billions of dollars on cybersecurity.

Policy and public awareness aside, cybersecurity experts are still in the early stages of wrestling with notions of a strategy to deal with cyber threats, something like the strategy that developed to deal with nuclear threats. Former CIA director David Petraeus argued that analyzing the possible flash points between two blocs of nuclear powers was relatively easy. Developing a strategy that takes potentially thousands of players into account is infinitely more complex.

When Keith Alexander retired as director of the NSA, he opened up offices for IronNet Cybersecurity, Inc., in downtown Washington. One of the imponderables in our system of government is how rapidly and easily secretaries of state, defense, and treasury, high-ranking military officers, congressmen, senators—indeed, even former presidents of the United States—are, upon departing office, able to transform their expertise, their experience, their contacts into extraordinarily high fees, contracts, and lucrative new businesses. Alexander's qualifications are unquestioned, but in a vibrant democracy, accusations of unseemly haste were inevitable.

That was certainly the thrust of a July 2014 article in the *Atlantic* titled "Keith Alexander's Unethical Get-Rich-Quick Plan?" Perhaps the question mark was intended to leave a shred of doubt, but a sense of outrage vibrated throughout the piece: "While responsible for countering cybersecurity threats to America," Conor Friedersdorf wrote, "Alexander presides over what he characterizes as staggering cyber-thefts and hugely worrisome security vulnerabilities. After many years, he retires.

And *immediately*, he has a dramatically better solution to this pressing national-security problem, one he never implemented in government but plans to patent and sell!"

Friedersdorf's implication was that Alexander is offering for vast profit a cybersecurity plan that he failed to develop while in the service of his country, and that he is utilizing the classified information that was once available to him. On the face of it, the suspicion and even the outrage are not unreasonable. But the actual answer is considerably less duplicitous and certainly less dramatic. At least—and not surprisingly—that is the impression Alexander gave when he sketched out the plan to me.

There are, as previously noted, more than three thousand electric power companies in the United States. Many of the smaller electric companies lack the resources and often the motivation to provide their operations with the best cybersecurity. Computer access to any one of them can provide access along the network to the SCADA and EMS systems that calibrate supply and demand for the grid as a whole. In theory, the government's most sophisticated intelligence gathering programs could monitor every single operation of every power company, every bank, every airline—in short, every critical industry in the country, alerting each industry to every incoming probe or attack. But to do so would be to violate every last vestige of Fourth Amendment or privacy protection.

What Alexander and his partner think they have found is an interim step—a commercial bridge between the critical industries and the NSA. Here's how Alexander described the plan: "If you take every power company network in the United States and you think of those like little saucers, and you put those saucers out on the table, some of those would overlap." If each of those networks had the capability to detect a cyberattack or exploitation, it would, in effect, provide a real-time map of the industry. Alexander's company would provide the necessary monitoring systems, and if these systems alerted an operating center that "somebody like the Iranians, for example, is scanning a number of those 'saucers' trying to find a weak spot," his company could then alert the government and other power companies to the possibility of "exploitation or attack."

Alexander is proposing something that would be the equivalent, he said, of a home protection service like ADT. For a monthly fee, ADT installs a burglar alarm system, cameras, fire detection equipment, and other devices of your choosing. "You don't have a security guy sitting in your basement and a cop and a fireman and all those," said Alexander. "You couldn't afford it." The way these home security systems work is that if there's a break-in, a gas leak, or a fire in your residence, sensors or cameras convey the information to a control center, which in turn calls the police or the fire department. In Keith Alexander's ideal world, his company would be running a cyber equivalent to the home security control center, monitoring the overlapping power company networks. He says, "That op center would see who's trying to get into any one of those, and then the issue is, can the op center share that with the government and with other entities in that sector?"

That may be where the plan implodes. "Right now," Alexander acknowledged, "you'd have a problem with the Electronic Communications Privacy Act for industry sharing with government, and for government sharing with industry."

As noted, this is not a good time to be arguing that the NSA needs more access to the security mechanisms protecting the

U.S. power industry. But the NSA has repeatedly insisted that it is not looking at the content of communications between and among Americans, and Alexander made the same argument for an operational control system that would scan the overlapping networks of U.S. power companies. Using Alexander's analogy to the ADT home security system, his company would merely be alerting the customer to the fact that an unauthorized party was trying to break into the house. The operative question is whether Alexander's company could then legally pass that information on to the government or another power company, which would be equivalent to ADT calling the police or alerting a neighbor. Detecting a cyber intrusion may not involve looking at content, but determining the nature of that intrusion almost certainly would. Once the police arrive on the scene, to complete the analogy, they would check out the house, and that's where privacy might be compromised. The advantage of the Alexander plan presumably lies in the fact that content would be monitored only when an intrusion has taken place.

Still, Alexander is a realist. He recognizes that his plan for a low-cost, home-protection-style alert system for small to medium-sized power companies will only function if the information can be shared across the industry and between industry and government. He also knows that a truly functional, top-to-bottom cybersecurity system for the electric power industry is not likely to happen until after a major, debilitating attack on the grid has occurred. Until then, "half of the Congress will say why we should do it, and then the other half will say why we shouldn't do it. And then they'll argue it, and they have no tactical understanding, most of them, about what they're arguing. Unless there's a true crisis, we're going to move slow." Even if everybody got behind it, said Alexander, "it would probably take five to seven years." At the moment, it would be fair to say that everybody is *not* behind the plan.

Gary Dylewski, a former fighter pilot and major general in the Air Force who in 2012 cofounded Patriot Solutions International with Ken Eichmann, a retired Air Force lieutenant general, also has a plan. His idea is to supply a secure backup supply of electricity for essential facilities in the event of a grid being taken down. Among the studies his company has been hired to undertake are a few energy security initiatives sponsored by the Department of Defense. The Defense Department, it turns out, is particularly worried about energy security on military bases.

Thirty-odd years ago, Dylewski explained, military bases typically produced their own power. That was prudent from a security point of view, but the operation of those power plants was also expensive. Gradually the Pentagon ordered all but a few of the bases to work out an arrangement with their local power company, plugging into the grid for efficiency and economy. "In today's world," Dylewski told me, "with the threats to energy security and threats to local grids, the military is now looking at how [we can] sort of go back to the future."

One of the answers currently being examined is nuclear plants. Not the 1,000–1,200 megawatt variety, the licensing of which became encumbered by so many safety considerations in the wake of Three Mile Island and Chernobyl that it's been decades since one was constructed. Rather, Dylewski is talking about using small, modular reactors of the kind used in certain naval ships over the past fifty years. Those reactors do not have to go through the licensing and certification process normally required by the Nuclear Regulatory Commission because they are considered to be on military property.

Patriot Solutions is working with the center for energy security at the University of Texas. Among their advisors is Dr. Dale Klein, who used to certify nuclear power plants for the NRC. He said that small nuclear reactors are demonstrably safe, given that they haven't had an accident during the fifty years they've been in use. Furthermore, Klein pointed out, if you put these reactors on military bases, you could greatly shorten the approval time for licensing and provide secure energy for the base. Since these modular reactors on military bases could produce more energy than they need, cooperative agreements could be worked out with local communities, providing emergency power to hospitals, police departments, and other first responders in the event that the grid goes down.

I asked Dylewski to lay out a best-case scenario: "If the Departments of Energy and Defense were to go to the president and say, 'We really think we ought to do this,' what are we looking at in terms of time?"

"If you used a technology that's familiar to the NRC, I would say between five and ten years," he told me. But since no one has yet authorized Gary Dylewski's plan, the clock hasn't even started.

It is not a hopeful scenario. One partial solution with promising implications is overshadowed by widespread and lingering concerns over the safety of nuclear power. The outline of a defensive structure against cyber intrusions on critical infrastructure runs afoul of privacy concerns and suspicions of an inappropriate scheme to cash in on an accumulation of classified background. Indeed, General Alexander's alma mater, the National Security Agency, is so proficient at what it does, that its virtuosity makes it suspect. There is no question that the most sophisticated technology, supported by the largest budgetary allocations and administered by the most capable electronic intelligence experts in government service, is at the National Security Agency. The NSA, said David Petraeus, is "far and away the most competent, capable, best-in-the-world entity" in terms of cybersecurity and analysis. Following the revelations made by Edward Snowden, it is also, Petraeus added, "a bit radioactive in terms of domestic cybersecurity." That sense of public mistrust has only been building since March of 2013, when James Clapper, the director of national intelligence, was asked at a congressional hearing whether the NSA collects "any type of data at all on millions or hundreds of millions of Americans."

"No, sir," said Clapper.

Senator Ron Wyden followed up. "It does not?"

"Not wittingly," said Clapper. "There are cases where they could inadvertently perhaps collect, but not wittingly." Documents released by Edward Snowden revealed that to be untrue. Clapper was subsequently obliged to make the embarrassing admission that what he had said was "erroneous" but the "least untruthful" answer he could give. By late spring of 2015 a congressional coalition of privacy advocates, libertarians, and Tea Party activists forced the NSA to surrender the bulk collection of phone records to the telephone companies themselves, requiring a court order for the NSA to gain access to those records.

General Petraeus was anticipating even further erosion of the NSA's authority. All this bad publicity may, according to

Petraeus, "accelerate the transition of some tasks that might have been performed by NSA . . . to the Homeland Security Department." It was clear he didn't think that was a good idea. "I don't think they have the same personnel rules. They certainly don't have the same cachet. I mean, people want to work for the NSA. They very much want to work for my old organization [the CIA]. I'm not sure the same could be said for the DHS yet." Petraeus was, if anything, understating the problem. Even some of Homeland Security's staunchest defenders acknowledge drawbacks. Larry Zevlin, who was until late 2014 director of the National Cyber and Communications Integration Center at DHS, told me he loved working at the department and thought the people were "phenomenal," but he conceded that the turnover at DHS was high. Janet Napolitano, who headed up the department for nearly five years, acknowledged the same point. If you were a scientist or mathematician working on cybersecurity, she told me, the NSA was simply the place to be.

It doesn't help that the Department of Homeland Security, whose mission, as declared on its website, is "a safer, more secure America, which is resilient against terrorism and other potential threats," ranks nineteenth out of nineteen large government agencies in an annual survey of best places to work in the federal government. DHS ranked nineteenth on effective leadership, nineteenth on empowerment, nineteenth on fairness, nineteenth on its senior leadership, nineteenth on supervisors, nineteenth on strategic management and teamwork, nineteenth on training and development. The intelligence community, including NSA, was ranked second (after NASA). The question then, is this: do we resolve the issue of domestic spying by taking aspects of cybersecurity out of the hands of the most competent agency and putting them in the hands of the least competent?

The Department of Homeland Security was created in an atmosphere of national trauma. The world's greatest superpower was made to realize its vulnerability to a handful of men armed with box cutters. Passenger planes could be reconfigured as missiles. We remain distracted to this day by the prospects of retail terrorism when we should be focused on the wholesale threat of cyber catastrophe. In such an event, the Department of Homeland Security would be working with industry to help them restore and maintain service. It should be focused on developing a more robust survival and recovery program for the general public; but DHS has neither the capacity to defend our national infrastructure against cyberattack nor the wherewithal with which to retaliate. A criminal attack would be the responsibility of the FBI; an attack on infrastructure by a nation-state or a terrorist entity would become the immediate responsibility of the Defense Department. Anticipating and tracking external cyber threats to U.S. infrastructure should be, by virtue of capability if nothing else, the responsibility of the NSA.

Limits that were established in a different era still exist on paper, but they are eroding in practice. The CIA is precluded, by law, from operating within the United States, but maintaining national boundaries in cyberspace may be impossible. Cyber Command is a military operation, tasked with organizing the defense of U.S. military networks. The extent to which it can participate in the defense of critical infrastructure within the United States remains murky, but sidelining critical U.S. defense capabilities because we haven't quite adapted to the notion that a major cyberattack can be as devastating as an invasion makes no sense. In practice, Keith Alexander explained, should a cyberattack be launched against U.S. infrastructure, the president gets his cabinet together, and the decision-making process is similar to what would happen in a nuclear command and control situation: "Because this is happening at network speed, there are things they should lay out ahead of time. . . . [I]f someone were coming in to take out the power grid . . . this could have a long-term impact on our nation." On that conference call with the president and his national security staff would be the secretary of defense, the chairman of the Joint Chiefs, and the commander of the U.S. Cyber Command. Homeland Security would be at the table, "because they have a role there to work with industry," but it is clear that Alexander regards their role in providing security as strictly secondary.

We have become disoriented by the similarities between the aftermath of a natural disaster and what will be required when it comes to helping the nation deal with the aftermath of a cyberattack on a grid. We need to adapt to the realization that at an as-yet-undetermined point a cyberattack on one of the nation's three electric power grids amounts to an attack on the United States. It would be no less an act of war than an air raid by enemy bombers or a strike by enemy missiles. When General Alexander describes the emergency cabinet meeting that the president would convene in such an event, he pointedly compares it to what would take place in a nuclear command and control situation. What would result directly from such an attack—the population flow, the extended distribution of emergency supplies, and the likelihood of civil unrest—would require the combined expertise and resources of many government agencies, but all would fall, inevitably, under the overall control and management of the military. It is the only organization with the equipment and manpower equal to the task. That will become all too self-evident after an electric power grid is disabled.

The imposition of order, the distribution of essential supplies, the establishment of shelters for the most vulnerable, the potential management of hundreds of thousands, if not millions, of domestic refugees will be complex enough if the general public knows what to expect and what to do. In the absence of any targeted preparation, in the absence of any serious civil defense campaign that acknowledges the likelihood of such an attack, predictable disorder will be compounded by a profound lack of information. It would be the ultimate irony if the most connected, the most media-saturated population in history failed to disseminate the most elementary survival plan until the power was out and it no longer had the capacity to do so.

20

....

Summing Up

Some wish for cyber safety, which they will not
get. Others wish for cyber order, which they
will not get.

—DAN GEER

We are at one of those evolutionary stages in history that tracks
the end of an era. It's not so much a hinge moment as a discern-
ible shift, a gradual transfer of control. The exercise of power,
transformational power, is popping up in unpredictable places
in unexpected hands. The Internet as a weapons system has
traditional applications for governments seeking to project
power, but its accessibility is not exclusive to nation-states. We
still need to worry about what the Russians, the Chinese, the
Iranians, and the North Koreans will do, and they need to be
equally concerned about us. But for the first time in the history
of warfare, small groups, even individuals, can undermine the
critical infrastructure of a state.

It was Ed Markey, then a congressman, who back in 2010

solicited the opinions of some of the nation's top national security experts on the vulnerability of the grid. He provided a redacted version of that confidential letter for this book. When I asked Markey to respond to officials at the Department of Homeland Security who insist that the grid is resilient, he said, "They are ignoring the warning of almost every national security expert who has studied the issue."

It is time to decide which experts we are prepared to trust. In researching this subject, I have found myself relying significantly on the expertise of George Cotter. His credentials, as former chief scientist at the National Security Agency, are a major factor, but at some point or another, all reporters find themselves confronting a moment of decision. Almost by definition, when we are dealing with complex subjects, we tend to be less knowledgeable than the sources we are interviewing. At one point or another in this process, each of us ends up trusting his gut—deciding, quite simply, how much confidence to place in each source. I think George Cotter knows what he's talking about. In April 2015 Cotter produced his fourth white paper in a series titled *Security in the North American Grid—A Nation at Risk*. He sends these white papers to policy makers and federal institutions charged with homeland defense. All the material cited is unclassified. Although the paper is technical, its conclusions are simple and stark:

> With adversaries' malware in the National Grid, the nation has little or no chance of withstanding a major cyberattack on the North American electrical system. Incredibly weak cybersecurity standards with a wide-open communications and network fabric virtually

guarantees success to major nation-states and competent hacktivists. This [electric power] industry is simply
unrealistic in believing in the resiliency of this Grid
subject to a sophisticated attack. *When such an attack
occurs, make no mistake, there will be major loss of life
and serious crippling of National Security capabilities.*
[Emphasis added.]

Cotter's voice is merely one of the most persistent and best-
informed, but his essential message hardly differs from the one
Leon Panetta, then secretary of defense, delivered to an audience of security executives in October 2012. Panetta warned
that an aggressor nation or extremist group could launch "a
destructive cyber-terrorist attack [that] could virtually para-
lyze the nation." Some of the potential threats Panetta cited
included the deliberate derailing of trains, the contamination
of urban water supplies, and "the shutdown of the power grid
across large parts of the country. The collective result of these
kinds of attacks could be a cyber Pearl Harbor."

Panetta was invoking the most vivid of World War II images, one that became instantly synonymous with a surprise
attack. But in the case of Pearl Harbor there was no question
as to the identity of the attacker. The very next day, President
Franklin D. Roosevelt declared war on Japan. We would have
no such immediate certainty in the event of a cyberattack. The
inability to quickly discover the identity of an aggressor undermines the threat of retaliation. Deliberate misdirection and the
chaos caused by the attack increases the possibility that a coun-
terstrike may be aimed at the wrong target. Neither the American public nor the international community has come to terms

yet with the notion that a major cyberattack would amount to an act of war, but a war that is as different from any previous war we have known as a nuclear conflict would be from conventional warfare. How do we prepare for something that we have not even adequately defined?

In an Oval Office conversation with *New York Times* columnist Thomas Friedman in April 2015, President Obama said, "Iran's defense budget is $30 billion. Our defense budget is closer to $600 billion. Iran understands that they cannot fight us." The president was talking to Friedman in the context of a nuclear agreement, stressing the disparity between any military force that Iran could hope to project and the U.S. military. Even so, the president certainly knew his statement to be only partly accurate. Iran surely understands that it cannot hope to wage a nuclear war with the United States and win, but Iran will continue pursuing its strategic interests by other means: terrorism, the use of surrogates, and, increasingly, cyber warfare. If it happens, when it happens, the size of Iran's military budget will be irrelevant. We may not even know with certainty that Iran launched the attack.

Shortly after taking over as secretary of defense in February 2015, Ashton Carter ordered the release of a thirty-three-page cybersecurity strategy. The document warns potential adversaries that they will suffer "unacceptable costs" if they conduct an attack on the United States. It's an interesting exercise in deliberate ambiguity, in that cyber intrusions—espionage, theft, distributed denial-of-service attacks against U.S. interests—are a daily occurrence. It's a bold warning, but the question of what constitutes a red line is left deliberately unclear. The document also acknowledges an urgent need "to reduce anonymity in cyberspace and increase confidence in attribution." Those two fragments from the Pentagon's cybersecurity strategy neatly summarize the dilemma: the United States is warning anyone who launches a cyberattack against this country of dire consequences, without defining what would trigger such a reprisal, while at the same time acknowledging that we also cannot be 100 percent certain of identifying the attacker.

In February 2015 President Obama, acutely aware of the need to synthesize the best cyber intelligence from all available sources, issued a presidential memorandum establishing the Cyber Threat Intelligence Integration Center (CTIIC), to be headed up by the director of national intelligence. Significantly, the president's memorandum directs that "indicators of malicious cyber activity and, as appropriate, related threat reporting contained in intelligence channels [be] downgraded to the lowest classification possible." This is intended to ensure the rapid sharing of intelligence material with the widest possible business and industry base. Downgrading its material to the lowest possible classification is not something that comes easily to the intelligence community, and it remains to be seen how enthusiastically the various intelligence agencies comply. Getting critical information from the government to private industry is only half the problem. Not only has private industry been reluctant to share information about cyber intrusions with government agencies, but it has actively lobbied against the proposition in Congress, with the result that cybersecurity legislation has languished in the Senate. Businesses have been concerned about word of successful hacking operations becoming public and hurting their image. They have been worried about news of electronic intrusions giving business rivals a competitive advantage. They have instinctively resisted ceding any of their independence and privileged information to government agencies. And, perhaps most of all, they've been nervous about loss of privacy and the potential legal consequences of permitting private information to become public. In the face of mounting cyber threats, businesses are slowly, painfully setting aside some of their concerns and considering a closer collaboration with federal agencies.

But apparently the fear of surrendering privacy to the U.S. government still looms larger than the threats posed by foreign governments and an array of unseen hackers. During the spring of 2015, the Senate was on the verge of passing legislation that would have given industry the right to "scrub" data, for privacy purposes, before handing information over to the government. The plan would have given the Department of Homeland Security the task of further sanitizing the information before conveying it to the NSA, or whichever government agency was deemed appropriate. It wasn't much of a plan. Indeed, it's difficult to imagine designing a procedure better suited to slowing down the sharing of critical information. Still, it was further than a coalition of privacy and pro-business advocates were willing to go. In mid-June of 2015, for the third year in a row, the measure was killed in the Senate.

It is fraught territory. There is probably no adjustment in our national mindset that will be more difficult to achieve than changing priorities in the tension between security and privacy. In early May 2015 a federal appeals court in New York ruled that the bulk collection of the American public's phone records is illegal. That was the news that got the headlines. The court's ruling, however, was more nuanced than the headlines

suggested. "Such expansive development of government reposi-
tories of formerly private records would be an unprecedented
contraction of the privacy expectations of all Americans," wrote
Judge Gerard E. Lynch. But then his ruling went on: "Perhaps
such a contraction is required by national security needs in the
face of the dangers of contemporary domestic and international
terrorism. But we would expect such a momentous decision to
be preceded by substantial debate, and expressed in unmistak-
able language."

Lynch is exactly right. The issue does require substantial
debate, and it is not easily resolved. There's nothing new about
the argument. It crops up every time there's a threat to national
security. The leaders of democracies have always argued that
they are operating at a disadvantage in their dealings with to-
talitarian governments. The Russians don't need to worry about
infringing on the privacy concerns of their citizens or their
critical industries. Neither do the Chinese. But it is in the nature of how democracies function that the pendulum of public
outrage can swing dramatically from one extreme to another,
depending on perceived threats to the national security, and
policy adjustments usually follow. As threat perceptions rise,
public concerns about privacy tend to diminish. In the immediate post-9/11 environment, very few concerns were raised about
the intrusiveness of law enforcement or intelligence agencies.
Indeed, criticism focused largely on the inability of the CIA,
NSA, and FBI to "connect the dots," on government's failure
to anticipate the attacks. As the years passed and the trauma of
9/11 receded in the public mind, the pendulum has swung back
dramatically on the side of protecting privacy and limiting the
invasiveness of our intelligence agencies.

What's different this time is that the very nature of cyber intrusion is a threat to privacy. Few members of Congress have invested more time in studying the tensions between cybersecurity and privacy than Senator Edward Markey (D-Mass.). On the one hand, he points out, you have the private sector—Google, Yahoo, and the like—engaged in gathering massive amounts of data on the individual habits of Americans so that they can sell that information to other vendors, who can then target their products to the most likely customers. "Well," said Markey, "what if a thirteen-year-old girl is Googling for information on anorexia? Should her mother be able to say, 'Stop re-marketing all that anorexia information onto my girl's computer. She's already sick. I want you to stop'? And the industry says, 'No, we don't want rules on that. You can't inhibit our ability to take the information of every individual American and re-market it to make money.'" But when the government expresses a similar interest in all of that information, not to make money but to protect the country, and approaches Google and Yahoo, "they'll say to government, 'No, we don't want to give up that information.'"

If we insist too adamantly on protecting privacy, we will sacrifice both free enterprise and security. In the age of the Internet, privacy is at risk no matter what we do. What's at issue is whether we are prepared to surrender some of our privacy to our own intelligence agencies in order to protect against even greater intrusions from a growing array of external enemies. Until the general public is made to understand the scope of the actual threat, the natural inclination will be to preserve what we know and value, against what we still suspect may never happen.

As things now stand, the general public would have a difficult time discriminating between an ordinary power outage and an act of war. No government agency has even laid the groundwork by establishing such a possibility in the public mind. Knowledge of that possibility is a necessary prerequisite if we are to have any hope of encouraging disaster preparation on the public's part. The implications of a weeks-long (let alone months-long) loss of electricity across large regions, especially those with significant urban populations, are sufficiently grim that at least a basic level of public awareness needs to be established.

"I think one of the lessons learned from the hurricane in New Orleans," David Petraeus told me, "is that if something like that happens, pull the trigger rapidly and get all hands on deck." Petraeus said that as a direct consequence of what happened after Hurricane Katrina, the military established liaison teams in all the states. He recommended that I talk with General Chuck Jacoby, also a four-star, who had just retired as commander of the U.S. Northern Command, which has the responsibility for homeland defense and military assistance in disaster relief throughout North America.

Among the challenges facing the NORTHCOM commander is the need to tread lightly while mustering the capacity to respond rapidly. There is in the United States a historical sensitivity toward the use of federal troops, particularly when it comes to maintaining or restoring order. "Every day," Jacoby told me, "I used to say that the NORTHCOM commander's job was to reconcile the will of the president with the authority of the governors. They own their state and they own their [National] Guard, and you know the power for authorities with enforcement capabilities really emanates from the people. So it comes up from the local police departments to sheriffs to the state to the Guard and then up to the president. And it's a very, very deliberate legal issue to use federal military forces in an armed capacity in the homeland."

It happens, but rarely. In 1957 President Eisenhower federalized Arkansas's entire National Guard and then reinforced the guard with units of the 101st Airborne Division when the state refused to integrate its schools. In 1968 the 82nd Airborne Division shipped a brigade from Viet Nam to Detroit to restore order after race riots broke out. In 1992 units of the 7th Infantry Division were dispatched to Los Angeles when rioting broke out in the wake of Rodney King's beating.

In anticipating the event of a power grid going down, however, the process will have to be streamlined and rehearsed. During the time that it takes to alert and dispatch military personnel and to mobilize the National Guard, local and state police will need to immediately secure the stores and warehouses containing essential supplies that will otherwise be stripped bare in a matter of hours. The authority exists, but without the regular conduct of combined exercises specifically designed to respond to the aftermath of a grid going down, critical supplies will be gone before law enforcement even arrives on the scene.

Jacoby was the beneficiary of lessons learned during and after Hurricane Katrina. The president now has the authority, at the request of a governor, "to direct the Department of Defense to commit resources for emergency work essential to preserve life and property in the immediate aftermath of an incident." In fact, Jacoby argued, disaster response in general has been vastly improved in the wake of Katrina "because there's a pretty good National Response Framework . . . and that's a team that knows how to support FEMA in a disaster." Again, though, the focus is almost always on natural disasters.

The question of maintaining security in the aftermath of a power grid being shut down, Jacoby added, can quickly be reduced to a matter of manpower. The U.S. military is a diminished force, with the army down to about 450,000 people. Whether that would be adequate, said Jacoby, is really problematic. NORTHCOM could come up with 50,000 or so troops fairly quickly, but then, said Jacoby, "if that's not sufficient then you have to start making choices between 'Am I sending that brigade to Iraq or am I sending that brigade to Afghanistan or am I making that deployment to Poland?'"

Jacoby seems a thoughtful man. He is torn between the discipline of military preparedness, with its indisputable value in a time of national crisis, and the American system, which is "designed," as Jacoby said, "for inhibiting federal abuse of power, specifically armed power in the homeland. And that's who we are as a people."

Jacoby is struck by the irony that while we have the most powerful means of communicating with the public that has ever existed, it will be essentially useless without electricity. The communicating needs to happen now. "This is all pre-disaster stuff that has to be done," he stressed.

From a purely domestic point of view, it should already be a settled issue how forces would be activated, and under whose command, as soon as the president is convinced that all or part of a power grid has been the target of a cyberattack. Maintaining public order and protecting the civilian population will become more difficult with each passing day. As FEMA administrator Craig Fugate acknowledged, it quickly becomes a matter of keeping as many people from dying as possible. It's food, potable water, and enough generators to keep water flowing and a waste disposal system functioning. There is not now an emergency food supply even remotely adequate to what the demand would be. Among what Jacoby described as the "pre-disaster stuff" that has to be figured out is a plan under which the federal government would acquire billions of dollars' worth of freeze-dried food, sufficient to feed tens of millions of people for a period of months. This alone will take years once the money is appropriated and the contracts have been signed.

Americans are accustomed to going where they want to go, when they want to go. Many city dwellers have focused their survival plans on just driving to the nearest state in which the power is still on. There is no guarantee that they will be invited to stay. To the contrary. One former state employee from a small rural state told me of strategy sessions planning how they would handle a mass evacuation from an affected city. Traffic police, state police, the National Guard, and civilian volunteers wearing official paraphernalia would be stationed in key locations, offering food, water, and directions to the next gas station. But the message was stark and simple: "Our state doesn't have the infrastructure to support large numbers of evacuees. Please keep moving." These are issues that are quietly being discussed on a state-by-state basis. There is no national strategy.

When one major sector of the country is without electricity and the rest of the country has power, what happens? Do states have the right and the legal authority to require domestic refugees, who have neither guaranteed shelter nor the funds to rent or buy shelter, to keep moving? What happens to the economy of the darkened states? With a diminished ability to generate revenue, how long will those states be able to count on the generosity of the rest of the country? Will the federal government establish refugee camps? Where? We have barely begun to consider the problems, let alone find the solutions.

Dan Geer is a computer analyst, admiringly described by colleagues as one of the industry's "thought leaders." Delivering the keynote address to the 2014 Black Hat hackers conference, Geer said the following: "Power exists to be used. Some wish for cyber safety, which they will not get. Others wish for cyber order, which they will not get. Some have the eye to discern cyber policies that are 'the least worst thing.' May they fill the vacuum of wishful thinking."

Rudy Giuliani looks back now on the events of 9/11 and their immediate aftermath, and invokes advice he received from a judge for whom he once clerked. The judge had told his young law clerk, "For every one hour in court, four hours of preparation."

"I think that point of relentless preparation is really important," Giuliani told me.

When September 11 happened, it was unanticipated. "My first reaction was," the former mayor said, "we're not prepared for this." New York City emergency personnel, he said, had engaged in relentless preparation with city officials, but "we hadn't gone through it in the context of airplanes being used as missiles attacking our buildings. We had thought of the fact of possible dirty bombs [or] a small nuclear attack." And yet, Giuliani maintained, the act of preparing itself was "enormously helpful." To this day, the former mayor believes that "the more you prepare, the better off you are going to be, even if you haven't quite anticipated the thing that happens."

The Virtue of a Plan

When the lights go on again all over the world.
—POPULAR SONG LYRICS FROM WORLD WAR II

It was in the days before the Internet, before social media, before satellite technology, before cellphones, before television. It was, needless to say, a very long time ago, and the world was at war.

Many of my earliest memories were crafted by images of World War II. In my father's arms, I would follow the arc of his finger against a late-evening summer sky, watching formations of Royal Air Force fighter planes heading toward the English Channel to engage the Luftwaffe. German bombers would be approaching the southern coast of England, accompanied by their protective escort of Messerschmitt and Heinkel fighter planes. My father would wake me a few hours later, as the RAF Hurricanes and Spitfires returned from their missions, and once again he would hold me and point to the sky, this time marking the gaps in their formations. I was two or three at the time, the unwitting repository of memories that I would only comprehend many years later.

In his magisterial volume *The Bombing War: Europe 1939–1945*, Richard Overy devoted an eye-opening chapter to what was, at least in 1940, an unprecedented episode in the history of warfare. "British society," he wrote, "was the first to be tested to see whether the fantastic images of social disintegration suggested in the air culture of the pre-war years would really be the outcome."

The British in 1940 were as innocent of what to expect from a massive German air offensive as Americans are today at the prospect of massive cyberattacks against key elements of the U.S. infrastructure. There was clear evidence that something nasty was brewing; exactly what form it would take was less clear. The world had been introduced only recently to the concept of a deliberate bombing campaign against civilian targets. On April 26, 1937, German pilots, members of the Condor Legion, flying in support of Francisco Franco's forces, infamously bombed the town of Guernica. There was no strategic value to the target. It was akin to a boy focusing the sun on an ant with a magnifying glass, an experiment to see just how much devastation aerial bombardment could inflict on a town.

That carnage, immortalized in Picasso's *Guernica*, would have been fresh in the minds of the British, but so would earlier reports of the Italian air force spraying mustard gas on civilian targets in Abyssinia in 1935 and 1936. Consider these events the emotional framework within which the German attacks on England were anticipated. The British government knew that it had to prepare, but for what, exactly, was unclear. Poison gas had been widely used during World War I. The Italian gas attacks in Abyssinia were another clue. It was hardly unreasonable to assume that Adolf Hitler, already busily engaged in the extermination of millions in specially constructed gas chambers, might order the use of poison gas against the British population.

The resulting civil defense planning was a strange mixture of thoughtful preparation and misplaced emphasis. Parliament had passed the Air Raid Precautions Act in 1937, which provided the lion's share of funding and an organizational structure that connected local and central government, but public bomb shelters were woefully inadequate. The great fear was poison gas. Thousands of decontamination chambers were set up, and nurses were trained to deal with the aftereffects of gas poisoning.

Great Britain declared war on Germany in September 1939, following the Nazi invasion of Poland; but Germany's massive bombing campaign against Britain did not begin until almost a year later. That year's grace period made a world of difference. The fact that Britain was a nation at war created an appropriate mindset; the delay provided the opportunity to prepare. "Between 1939 and 1940," wrote Overy, "an army of regulars and volunteers was created capable of manning the front line; for the rest of the civil population habits of obedience to the blackout regulations, gas-mask drills, air-raid alerts and evacuation imposed on everyone an exceptional pattern of wartime behavior that persisted until the very end of the war."

Both of my parents were refugees from Germany, denied British citizenship until the war was over. Until that time, my father was also denied permission to work. Age and his ambiguous national status made service in the British armed forces impossible, but he could, and did, serve in the Home Guard. And there begins my first, vague awareness of civil defense. As best I can recall, my father and a neighbor, also a volunteer in the Home Guard, would patrol the neighborhood after dark armed with a long-handled whisk broom and the metal cover of a rubbish bin. It is not a heroic image, but in the catalogue of civil defense measures implemented by the British government, even these teams may have been useful. Beyond the practical impact of saving the occasional home from fire, such programs gave participating civilians a feeling of usefulness, a sense of connection with a larger mission.

The Battle of Britain raged in the skies over London and other English cities during 1940 and 1941. That was the worst of the aerial bombardment, but even afterward German bombers carried out missions intent on destroying England's infrastructure and terrorizing its civilian population. Among their weapons was the incendiary bomb, a small, finned tube filled with magnesium and carbide. It would explode on impact with an intense flame that burned for up to fifteen minutes. Left alone, incendiary bombs were deadly and highly destructive, but there were relatively simple measures that were sometimes enough to counteract them. An instructional film shown at civil defense training sessions and before the feature presentation at local cinemas urged audiences to acquire hand pumps. One person would immerse the pump in a bucket of water, pushing a plunger up and down, while another member of the family was shown directing the stream onto a burning incendiary bomb.

That was the high-tech approach. My father and his partner had been instructed to apply what might have been called the "sweep and smother" method. If an incendiary device landed on top of a home, it would in short order burn through the roof and set the house on fire. The broom man on my father's team had been instructed to climb onto the roof and sweep the bomb to the ground, where his rubbish lid cohort would smother the thing. Whether my father ever had the occasion to apply this technique, I do not know. That, however, was the plan, and a plan can be a virtue in and of itself.

We had a small bomb shelter in our garden. The owner of the house must have had it installed before the war, and I believe we used it only once. It was cold and damp, and my good German mother, who rarely asserted herself, said she would rather take her chances in the house. My father had acquired a heavy metal desk, and throughout the war I slept under that desk, protected, if not from a direct hit, at least from the danger of flying glass and shrapnel.

With the clarity of hindsight we can apply clinical analysis to what worked, what didn't, and why. That bomb shelter of ours at 19 Haslemere Gardens was an Anderson shelter, a cheap construct of curved metal panels installed several feet deep in the soil of our backyard. These were, according to Overy, "not even remotely bombproof." My mother would have felt quietly vindicated.

There were, the government concluded, only two viable options: shelter and evacuation. By March 1940, major cities contained shelter space for almost half of their population of 27.6 million, but this was less impressive than it sounds. Thirty-nine percent of domestic shelters were regarded as likely ineffective. Nor were the public shelters much better, and Prime Minister Winston Churchill and his cabinet were initially disinclined to devote the resources or manpower that would have been necessary to improve their quality. So the emphasis was placed on evacuation. On September 1, 1939, Overy noted, 1,473,500 people—children, pregnant women, new mothers, the disabled—left England's cities for the comparative safety of the countryside, where they would be rehoused. Despite these huge initial numbers, it turned out to be a highly unpopular option, particularly in light of the fact that German bombing had not yet begun. By January 1940, wrote Overy, around 900,000 of the evacuees had returned to their urban homes. Once the bombing did begin, evacuation was no longer an option. "On 7 September [1941], the first day and night of heavy bombing in London, several thousand Londoners bought tickets for the Underground and stayed put in the stations and tunnels." During the early weeks of the bombing more than 120,000 Londoners took shelter in that fashion. Around 65,000 stayed there even through the winter, though the bombing declined. The platforms and tunnels had no toilet facilities and they grew increasingly filthy. There was nothing to eat or drink, nor had anyone thought to provide cots for sleeping.

There had never been a sustained bombing campaign against an urban target. With no precedent to guide its conclusion, the British government simply assumed that most German bombing would occur during daytime, so no preparations had been made to turn public or domestic shelters into dormitories. In the final analysis, the vast majority of Londoners opted for neither evacuation nor the bomb shelters. People took their chances at home.

For all that, there was an organized structure in place. There was a level of civilian discipline that served the country well

when the horrendous bombing finally began. Churchill's elo-
quence and the royal family's refusal to evacuate from Bucking-
ham Palace reinforced the population's self-image of enduring
hardship with a stiff upper lip. The British were emotionally
prepared for whatever might come, even if what ultimately
came was not what had been expected.

During the latter years of the war the Germans unveiled a
new class of weapons to be deployed against civilian targets: the
so-called *Vergeltungswaffen,* or retaliatory weapons. We children knew them as "doodlebugs" or "buzz bombs." They were
invented and designed by scientists, some of whom (depending
on where and by whom they were captured) would, after the
war, end up in either the U.S. or Soviet space programs. The
V-1 was a precursor of the cruise missile; the V-2 was an early
version of the liquid-fueled ballistic missile. We were unwitting
witnesses to the dawn of a new age. All I knew as a child was
that doodlebugs emitted a high-pitched whistle as they hurtled
through the sky, and then, when they were over their targets,
the whistle would stop as the missile fell to earth. There would
be a brief period, perhaps ten seconds of silence, before impact
and the sound of an explosion. Nothing would have protected
our family against a direct hit, but I didn't know that. I knew
that when I heard the whistle of a V-1 that I was to race for my
father's study and duck under his desk. It was a useful concept,
not just as a practical matter but also for the confidence it in-
stilled in a child who regarded the whole exercise as something
of a lark.

One additional memory from my childhood, which may have
been formed precisely on May 8, 1945: VE Day. It was after dark
when my father took me out onto the street to show me a sight I

had never seen before: lighted street lamps. My entire life until that moment had been lived under what we knew as the "blackout." Every window in every building throughout England had some variation of a blackout shade, intended to keep even glimmers of light from serving as beacons to German bombardiers. There was a popular song at the time: "When the Lights Go On Again (All Over the World)." For me, they weren't going on again; they were being lighted for the very first time. Much of Britain's civilian population had faced infinitely more harrowing circumstances than ours. What lingers, after all these years, is the sense of preparedness, of having a plan, of being ready for whatever might come.

In a sense, preparing for the unknown has always been the challenge facing civil defense planners. How does a country's leadership draw the appropriate line between prudence and paranoia when neither the timing nor the exact nature of a threat to national security can be defined? That was the question confronting the United States in the years following the Soviet Union's first successful test of an atomic bomb in 1949. Like the British ten years earlier, American civil defense planners concluded that their options essentially boiled down to shelters and evacuation.

General Dwight D. Eisenhower, who masterminded the D-Day invasion as Supreme Allied Commander, had been impressed by the quality of Germany's famous highway system, the autobahn. In 1956 President Eisenhower succeeded in ramming legislation through Congress to approve the U.S.

Interstate Highway System. Eisenhower, the former military commander, envisioned the easier movement of troops and matériel in times of national crisis and the massive evacuation of cities in the event of an atomic attack. What that highway system produced, perhaps unintentionally, was almost immediate social and economic transformation. It led to, among other things, the vast exodus of industry and businesses from America's cities, which in turn encouraged the development of suburbs, with their shopping centers and networks of gas stations, and a new category of Americans—commuters. Not that Eisenhower's fears of a nuclear attack seemed unjustified in the 1950s. Indeed, a number of communities around the country ran vast evacuation exercises.

One such was Binghamton, New York. On Sunday, May 5, 1957, 1,448 men, women, and children from Binghamton took part in an exercise code-named Operation Evac-12. It would take them from the Twelfth Ward on the city's east side to the appropriately named village of Deposit, some twenty-eight miles away. A contemporaneous account in a local newspaper, the *Binghamton Press,* reported that this would be "the first mass evacuation of such magnitude in the nation." The newspaper went on to report: "At 1:30, the first of about 500 people emulating panic left their homes and climbed into about 120 automobiles. Others weren't displaying panic, as planned[,] got into automobiles, and all began their 28-mile trip along State Route 17. Nearly 110 people boarded an Erie-Lackawanna passenger train, which served as a traveling hospital for patients. School buses were also deployed. Had this been a real attack, Deposit would've been expected to handle upward of 10,500 people."

The exercise was a one-day affair. It lasted only until that same evening, when the village of Deposit hosted about three thousand evacuees, civil defense personnel, and observers for a dinner of chicken, biscuits, coffee, and doughnuts. Our reporter of that day's events apparently did not ask the villagers how they would have handled the flood of "upward of 10,500 people" expected in the aftermath of an actual atomic bomb attack, nor did anyone speculate on how Deposit would have managed such a horde through a stay of indefinite duration.

What mattered at the time was not so much the likelihood of actual survival as the perception of a ready-for-anything level of preparedness. Michael Chertoff, who served as secretary of homeland security during the administration of George W. Bush, told me that civil defense planners in the 1950s may not have been as naive as they seemed. They were under no more illusions about the feasibility of mass evacuations than they were about the effectiveness of "duck and cover" drills in preparing for an atomic attack. These types of exercises, said Chertoff, were largely intended to convey to the Soviet Union the impression that the United States was determined to defend Western Europe even at the risk of all-out war.

That was during the brief age of the atomic bomb. When the even more deadly hydrogen bomb became a reality, it seemed that the government grew more invested in optimizing survival plans. In 1956 Congressman Chester E. Holifield, whose considerable clout derived from his chairmanship of the House Committee on Government Operations, publicly stated that there wouldn't be time for evacuation and that exercises such as "duck and cover" were little more than a charade. He chastised the Federal Civil Defense Administration (FCDA) for proposing these "cheap substitutes for atomic shelter." The FCDA ended the debate by calling the chairman's bluff, proposing a distinctly non-cheap, $32 billion program providing tax incentives and special low mortgage rates to households that built or included shelters. Again, this was 1956, when a $32 billion federal program would have been considered exorbitantly expensive. President Eisenhower and Congress compromised by building a classified underground bunker for members of Congress beneath the Greenbrier Resort in West Virginia.

Once the impact of a hydrogen bomb explosion was fully understood by the public, however, even New York City's ubiquitous bomb shelters, stocked as they were with blankets, biscuits, and water, were revealed as largely ineffective. It had all been part of a campaign to convey to the Soviets Washington's confidence in the ability of American cities to survive the impact of a nuclear explosion—and, therefore, the firmness of the U.S. commitment to defend Europe against a Soviet attack.

Looking back, it's a little difficult to determine where efforts to reassure the American public began and the campaign to mislead Moscow ended. Exercises were widespread and conducted with absolute seriousness. In July 1957 mock atomic bombs were dropped on a hundred American cities, and Boy Scouts were assigned to work with Civil Defense, searching for lost individuals, administering first aid, rescuing people and protecting animals.

Nebraska, home to Strategic Air Command outside Omaha and to a number of nuclear missile silos, had always been considered a prime Soviet target. Thinking about the unthinkable in 1963 included a bizarre exercise that involved a two-week survival test for thirty-five cows, one bull, and two student cow-hands. History records that Dennis DeFrain and Ike Anderson cared for the test herd, buried under five feet of dirt and provisioned with cattle feed and water in a 10,000 gallon tank. There was an auxiliary generator that would be "available if electrical power was interrupted by an atomic blast." While it's easy now to ridicule some of Nebraska's civil defense experiments, it may in fact have been the realization that sheltering a couple of million head of livestock—or finding durable shelter for, or evacuating, tens of millions of people—was impractical, to put it gently, that stimulated alternative thinking in Moscow and Washington. Ultimately, such exercises in speculation led to the evolution of a strategy that abandoned both shelter and evacuation as viable protection for millions of civilians. The new strategy was based on the proposition that as long as the United States and the Soviet Union could maintain a reasonable equilibrium of terror, as long as either side retained the capacity to respond to a nuclear attack with equal or greater force, there would be a disincentive to launch a surprise first strike.

This tenuous but crucial détente was possible only because of the deep awareness both nations had cultivated of the threat of nuclear war. Both sides needed to go through the motions to reach the conclusion that civil defense planning was essentially useless in the context of a nuclear war, which in turn led to the relative safe harbor of strategic arms limitation. Likewise, seventy-five years ago, the reality of an unprecedented campaign of aerial bombardment against its civilian population obliged the British government to rethink its earliest concept of civil defense. The initial planning had been imperfect, the focus was often misdirected, but there were preparations, which gave the British a sense of direction amid chaos and without which the destruction wreaked by the Luftwaffe would have been even more devastating.

There is, as yet, no real sense of alarm attached to the prospect of cyber war. The initial probes—into our banks and credit card companies, into newspapers and government agencies—have tended to leave us unmoved. The public was engaged by the North Korean hacking attack on Sony, with the saucy, gossipy tidbits capturing attention for a few weeks, but there was no real sense of outrage or danger. In reality, the ranks of our potential enemies have never been this deep. Our points of vulnerable access are greater than in all of previous human history, yet we have barely begun to focus on the actual danger that cyber warfare presents to our national infrastructure. Past experience in preparing for the unexpected teaches us that, more often than not, we get it wrong. It also teaches that there is value in the act of searching for answers. Acknowledging ignorance is often the first step toward finding a solution. The next step entails identifying the problem. Here it is: for the first time in the history of warfare, governments need to worry about force projection by individual laptop. Those charged with restoring the nation after such an attack will have to come to terms with the notion that the Internet, among its many, many virtues, is also a weapon of mass destruction.

NOTES

....

All quotes attributed in the text but not cited below derive from personal interviews.

PART I: A CYBERATTACK

CHAPTER 1: **Warfare 2.0**

7 **"all 2.1 million current federal employees":** David E. Sanger and Julie Hirschfeld Davis, *New York Times* "Data Breach Tied to China Hit Millions," June 5, 2015

8 **"could disrupt air traffic control operations":** Government Accountability Office, *Information Security: FAA Needs to Address Weaknesses in Air Traffic Control Systems,* GAO-15-221, March 2, 2015, 13.

9 **And the United States, in collaboration with Israel:** David Sanger, *Confront and Conceal: Obama's Secret Wars and Surprising Use of American Power* (New York: Random House, 2012).

9 **Iran wasted little time:** Based on conversations with U.S. military intelligence sources who spoke on the condition of anonymity.

9 **Russia, China, and Iran, among others:** Ibid., and also from conversations with George R. Cotter, former chief scientist for the National Security Agency.

12 **the British market research company YouGov:** "Poll Results: Barack Obama Birth," YouGov/Economist, February 12, 2014.

12 **And in 2013 the Pew Research Center:** "Climate Change and Financial Instability Seen as Top Global Threats," Pew Research Center, June 24, 2013.

12 **One year later, another Pew Research Center survey:** Seth Motel, "Polls show most Americans believe in climate change, but give it low priority," Pew Research Center, September 23, 2014.

14 **the Internet is also giving rise to "filter bubbles":** Eli Pariser, *The Filter Bubble: What the Internet Is Hiding From You* (New York: Penguin Press, 2011).

15 **sent a confidential letter, not previously released:** A congressional source provided a copy of the letter.

CHAPTER 2: **AK-47s and EMPs**

17 **"This wasn't an incident where Billy-Bob":** Rebecca Smith, "Assault on California Power Station Raises Alarm on Potential for Terrorism," *Wall Street Journal*, February 5, 2014.

19 **In a conversation almost two years:** I encountered considerable reluctance within the electric power industry to discuss vulnerabilities of the grid. When one senior executive from a major power company did finally agree to talk, he would do so only on the condition that neither his name nor that of his company be used. Additional industry viewpoints were reflected in a conversation with two executives from Edison Electric Institute.

20 **In early April 2015, the Pentagon:** "U.S. Aerospace Command Moving Comms Gear Back to Cold War Bunker," Yahoo News, April 7, 2015.

21 **Admiral William Gortney, who in December 2014 took command:** Gortney provided this EMP explanation at a Pentagon news briefing on April 7, 2015.

22 **The commission's report, available online:** Report of the Commission to Assess the Threat to the United States from Electromagnetic Pulse (EMP) Attack, April 2008, available at www .empcommission.org.

22 **who warned of the growing threat of an EMP attack:** James Woolsey and Peter Vincent Pry, "The Growing Threat From an EMP Attack," *Wall Street Journal*, August 12, 2014.

CHAPTER 3: **Regulation Gridlock**

25 **One hot summer afternoon in August 2003:** J. R. Minkel, "The 2003 Northeast Blackout—Five Years Later," *Scientific American*, August 13, 2008.

26 **A U.S.-Canadian task force:** U.S.-Canada Power System Outage Task Force, *Final Report on the August 14, 2003 Blackout in the United States and Canada: Causes and Recommendations*, April 2004.

26 **The U.S.-Canadian task force recommended the implementation:** U.S.-Canada Power System Outage Task Force, *Final Report on the Implementation of the Task Force Recommendations— Natural Resources Canada, U.S. Department of Energy*, September 2006, 46.

29 **The grid has been operating according to reliability standards since:** To help me navigate the arcane world of grid reliability standards and the evolving relationship between NERC, representing the power industry, and FERC, which represents the federal government in its dealings with the industry, I engaged the services of Ryan Ellis. Dr. Ellis is a postdoctoral fellow at the Belfer Center for Science and International Affairs at Harvard's Kennedy School. I sent transcripts of key interviews and rough drafts of relevant chapters to Dr. Ellis for his review and comments. In his defense, let me say that I did not always follow his advice.

30 **NERC has an enforcement capacity as well:** The figures are pulled from an investigation by USA Today and 10 other Gannett newspapers and television stations across the country into the security of the power grid. Steve Reilly, "Bracing for a Big Power Grid Attack: 'One Is Too Many,' " *USA Today*, March 24, 2015.

CHAPTER 4: **Attack Surfaces**

36 **"So it's winter," said Clarke:** I conducted two lengthy interviews with Richard A. Clarke and am further indebted to Clarke for his excellent book *Cyber War: The Next Threat to National Security and What to Do About It,* which he coauthored with Robert Knake. It remains, as it was when first published in 2010, a first-rate and eminently readable primer on the subject and was invaluable in introducing this reporter to the subject.

40 **"also secretly recorded what normal operations at the nuclear plant":** William J. Broad, John Markoff, and David E. Sanger, "Israeli Test on Worm Called Crucial in Iran Nuclear Delay," *New York Times,* January 15, 2011. David Sanger deserves a separate endnote, in that his reporting on cybersecurity issues for the *New York Times* has been consistently excellent. He has been particularly generous in sharing his expertise and some of his sources with me.

41 **When Thomas Eric Duncan, infected with the Ebola virus:** Kevin Sack, "Downfall for Hospital Where Ebola Spread," *New York Times,* October 15, 2014.

42 **Whenever Homeland Security or the Federal Energy Regulatory Commission has hired:** This and the remainder of the information in this paragraph were provided by Richard Clarke.

43 **The problem with air-gapping:** Memo from Dr. Ryan Ellis to the author.

CHAPTER 5: **Guardians of the Grid**

46 **Indeed, in 2013 President Obama issued an executive order:** White House Office of the Press Secretary, "Executive Order—Improving Critical Infrastructure Cybersecurity," February 12, 2013.

50 **Among the experts I've consulted who now provide advice on cybersecurity:** Tom Ridge and Michael Chertoff are former secretaries of homeland security, Richard Clarke and Howard Schmidt were White House advisors on cybersecurity, Major General Brett Williams was director of operations at Cyber Command, and General Keith Alexander was director of the NSA.

CHAPTER 6: What Are the Odds?

58 **Leon Panetta told an audience of security executives:** Panetta delivered his remarks to members of the Business Executives for National Security aboard the Intrepid Sea, Air and Space Museum in New York, October 10, 2012.

59 **Quoted in a comprehensive *Washington Post* article:** May 31, 2015 by Craig Timberg.

60 **some parliamentarians in Germany:** Chris Miller, "Germany–U.S. Spy Scandal: Typewriters & Intelligence," *Cicero Magazine*, July 18, 2014.

60 **According to one military cyber specialist:** Several military cyber specialists who are still on active service spoke with me on the condition that neither their identity nor their rank or branch of service would be revealed.

61 **A glance at one of those digital attack maps online:** Just typing "DDoS" into any search engine will provide several visual representations of a digital attack.

64 **Ajit Jain is the CEO of the Berkshire Hathaway Reinsurance Group:** When I approached Warren Buffett to discuss the subject of insuring against cyberattacks, he recommended talking to Ajit Jain.

CHAPTER 7: Preparing the Battlefield

70 **writing for *Wired* magazine:** James Bamford, "The NSA Is Building the Country's Biggest Spy Center (Watch What You Say)," *Wired*, March 13, 2012.

72 **the *New York Times* reported that a Russian crime ring:** Nicole Perlroth and David Gelles, "Russian Hackers Amass Over a Billion Internet Passwords," *New York Times*, August 5, 2014.

73 **ten American financial institutions were revealed:** Matthew Goldstein, Nicole Perlroth, and David E. Sanger, "Hackers' Attack Cracked 10 Financial Firms in Major Assault," *New York Times*, October 3, 2014.

73 **Only days later, reports surfaced that Russian hackers had exploited:** Mark Scott, "Russian Hackers Used Bug in Microsoft

Windows for Spying, Report Says," *New York Times*, October 14, 2014.

75 **A U.S. security firm, Crowdstrike, spent much of 2014:** Nicole Perlroth, "Report Says Cyberattacks Originated Inside Iran," *New York Times*, December 2, 2014.

75 **"Iranian hackers are trying to identify computer systems":** Frederick W. Kagan and Tommy Stiansen, "The Growing Cyberthreat from Iran: The Initial Report of Project Pistachio Harvest," American Enterprise Institute Critical Threats Project and Norse Corporation, April 2015.

77 **described the handicap as being denied "home field advantage":** Michael Hayden, "American Intelligence and the 'High Noon' Scenario," *Wall Street Journal*, October 30, 2013.

CHAPTER 8: **Independent Actors**

84 **The *New York Times* and the *Wall Street Journal* ran lengthy investigations:** Michael Cieply and Brooks Barnes, "Sony Cyberattack, First a Nuisance, Swiftly Grew Into a Firestorm," *New York Times*, December 30, 2014; Ben Fritz, Danny Yadron, and Erich Schwartzel, "Behind the Scenes at Sony as Hacking Crisis Unfolded," *Wall Street Journal*, December 30, 2014.

86 **the total number of Internet protocol addresses in North Korea:** Nicole Perlroth and David E. Sanger, "North Korea Loses Its Link to the Internet," *New York Times*, December 23, 2014.

87 **they turned out to be "largely symbolic":** David E. Sanger and Michael S. Schmidt, "More Sanctions on North Korea After Sony Case," *New York Times*, January 2, 2015.

87 **North Korea has artillery in abundance:** Erik Sofge, "Can North Korea Really 'Flatten' Seoul?" *Popular Mechanics*, November 24, 2010.

88 **"to close a chapter from the first age of mass destruction":** Kim Zetter, *Countdown to Zero Day* (New York: Crown, 2014), 375.

NOTES

PART II: A NATION UNPREPARED

CHAPTER 9: Step Up, Step Down

94 **The Department of Energy reports:** This Department of Energy paper simultaneously claims that there may be as many as tens of thousands of large power transformers, while acknowledging that since the industry won't reveal the exact numbers, it has to speculate on the precise number. The claim that the United States has a larger base of installed LPTs than any other country is on page vi of the report. Infrastructure Security and Energy Restoration Office of Electricity Delivery and Energy Reliability, U.S. Department of Energy, *Large Power Transformers and the U.S. Electric Grid,* April 2014.

94 **"Should several of these units fail at the same time":** Ibid., 1.

95 **"Because LPTs are very expensive":** Ibid., 5.

97 **ran slightly more than thirteen minutes:** My interview with Secretary Johnson was taped and the section in question was timed.

98 **Executives from the Edison Electric Institute:** As mentioned elsewhere in the book, Edison is the association for shareholder-owned electric companies, which comprise the majority of the industry.

CHAPTER 11: State of Emergency

126 **There were reports in the wake of Hurricane Katrina:** Kevin Johnson, "Katrina Made Police Choose Between Duty and Loved Ones," *USA Today,* February 21, 2006.

127 **the financial cost of the wars in Afghanistan and Iraq:** Amy Belasco, "The Cost of Iraq: Afghanistan, and Other Global War on Terror Operations Since 9/11," Congressional Research Service, December 8, 2014.

127 **The Transportation Security Administration:** "The Transportation Security Administration and the Aviation-Security Fee," House Budget Committee Publications, December 10, 2013.

CHAPTER 12: **Press Six If You've Been Affected by a Disaster**

132 **One Red Cross website lists:** "Ways to Donate," American Red Cross, www.redcross.org.

132 **journalists from Pro Publica and National Public Radio:** Justin Elliott, Jesse Eisinger, and Laura Sullivan (NPR), "The Red Cross' Secret Disaster," ProPublica and NPR, October 29, 2014.

133 **Not surprisingly, the article drew an immediate and angry response:** Laura Howe, "American Red Cross Responds to Inaccuracies in ProPublica and NPR Stories," *American Red Cross Blog*, RedCrossChat.org, October 29, 2014.

PART III: SURVIVING THE AFTERMATH

CHAPTER 13: **The Ark Builders**

141 **Where details are provided, they stagger the imagination:** Genesis 5:32–10:1.

141 **Rashi, the eleventh-century French rabbi:** Judith Frishman and Lucas Van Rompay, *The Book of Genesis in Jewish and Oriental Christian Interpretation* (Lovanii: Peeters, 1997), 62–65.

142 **National Geographic channel broadcast a docudrama:** *American Blackout*, directed by Jonathan Rudd (National Geographic, 2013), DVD.

143 **The show's first program of its second season:** "Series Premiere of 'Doomsday Preppers' Launches National Geographic to Its Highest-Rated Tuesday Night Telecast Ever," *The Futon Critic*, February 9, 2012.

CHAPTER 14: **Some Men *Are* an Island**

155 **That little nugget was tucked away:** Liz Moyer, "For Sale: Renovated Luxury Condo; Can Survive Nuclear Attack," *Wall Street Journal*, November 9, 2014.

156 **"What is it, a thousand miles?":** According to www.distance-cities.com, it is 837 miles by car. Presumably the distance by bi-

cycle would be similar, although you might want to steer clear of major highways.

159 **Not all members of the family are convinced:** One of their grown daughters, Craig explained, has an aversion to guns and would not bring her family to the retreat in an emergency.

CHAPTER 15: **Where the Buffalo Roamed**

169 **Not only would he and his ranch hands survive a cyberattack:** As noted, Bob Model is in his early seventies. When I asked whether he took any prescription medicines and how he would maintain that supply, Bob conceded that this was a problem he had not considered.

170 **The theme of *Bowling Alone*:** Robert D. Putnam, *Bowling Alone* (New York: Simon & Schuster, 2000).

171 **The website for the *Wyoming Tribune Eagle*:** The list can be found at www.wyomingnews.com/clubs-and-orgs.

173 **statistics from the Centers for Disease Control (CDC) tell a different story:** The CDC's National Center for Health Statistics compiles its database from death certificates filed in each state and based on reports from attending physicians, medical examiners, and coroners.

175 **this is a place that has never been put to the test:** Unless, of course, you count the 10,700 Japanese Americans who were interned at the Heart Mountain Relocation Center during World War II. It's an option for accommodating large numbers of urban refugees, but not an appealing one.

CHAPTER 16: **The Mormons**

180 **I would be receiving a call from Henry Eyring:** The three top leaders of the Church of Jesus Christ of Latter-day Saints occupy what is known as the First Presidency. Thomas S. Monson is the president; Henry B. Eyring holds the post of First Counselor, putting him next in line.

181 **Joseph Smith's translation of the Book of Mormon:** To the extent that these next three chapters recount Mormon history or belief sys-

tem, they do so almost exclusively through the prism of the church itself. Richard Turley, assistant church historian, gave generously of his time in helping me understand some of the underlying events that shaped the Church's emphasis on disaster preparation.

185 **"If you are without bread, how much wisdom can you boast":** Brigham Young, *Journal of Discourses,* 8:68.

CHAPTER 17: **State of Deseret**
187 **"Have you ever paused to realize":** Ezra Taft Benson, "Prepare for the Days of Tribulation," *Council of the Twelve,* October 1980.

190 **and then the Second Quorum of the Seventy:** For all the church's strict hierarchy, there seems at times to be an almost whimsical bent toward titles and designations that may bewilder outsiders: "bishops" at all levels of the church hierarchy; "elders," most of whom are under twenty-five; and "quorums" that may or may not have their numerically designated membership. While the Quorum of the Twelve Apostles does, indeed, have twelve members, the Quorums of Seventy are less constrained by their labels. At Church headquarters in Salt Lake City, the First Quorum of the Seventy is, I was told, currently made up of 314 members.

191 **In a 2007 article for *Mother Jones*:** Stephanie Mencimer, "Mormons to the Rescue," *Mother Jones,* December 28, 2007.

CHAPTER 18: **Constructive Ambiguity**
201 **It has taken the church a long time:** "On September 11, 1857, a band of Mormon militia, under a flag of truce, lured unarmed members of a party of emigrants from their fortified encampment and, with the Paiute allies, killed them. More than 120 men, women and children perished in the slaughter." From the flyleaf of *Massacre at Mountain Meadows,* by Ronald W. Walker, Richard E. Turley Jr., and Glen M. Leonard (Oxford: Oxford University Press, 2008). For many years, the Mormon Church denied any culpability in the massacre. What's significant about this book is that Richard Turley, one of the coauthors, is assistant church historian, while Glen Leonard is former director of

church history and art. The book, in other words, was sanctioned by the church.

CHAPTER 19: Solutions

211 **"proposed a government-industry cyber war council"**: Carter Dougherty, "Banks Dreading Computer Hacks Call Cyber War Council," Bloomberg, July 8, 2014.

212 **That was certainly the thrust**: Conor Friedersdorf, "Keith Alexander's Unethical Get-Rich-Quick Plan," *Atlantic,* July 31, 2014.

216 **who in 2012 cofounded Patriot Solutions:** The Patriot Solutions website describes the Colorado Springs–based company as focusing on energy security, engineering services, procurement, and logistics.

218 **"Not wittingly":** In an interview with NBC's Andrea Mitchell, he said that "I responded in what I thought was the most truthful, or least untruthful manner, by saying no," though he also called his answer "too cute by half." He indicated that his response to Wyden turned on a definition of "collect:" "There are honest differences on the semantics of what—when someone says 'collection' to me, that has a specific meaning, which may have a different meaning to him."

219 **It doesn't help that the Department of Homeland Security:** The "Best Places to Work in the Federal Government" rankings are produced by the Partnership for Public Service, a nonprofit organization that seeks to strengthen the federal civil service.

CHAPTER 20: Summing Up

224 **Cotter produced his fourth white paper:** Cotter sends his white papers to a selected list of recipients, but does not make them available to the public.

225 **Panetta warned that an aggressor nation or extremist group:** Transcript, U.S. Department of Defense, October 11, 2012.

226 **In an Oval Office conversation:** Thomas Friedman, "Iran and the Obama Doctrine," *New York Times,* April 5, 2015.

226 **Ashton Carter ordered the release:** U.S. Department of Defense, *DOD Cyber Strategy,* April 2015.

227 **issued a presidential memorandum:** White House Office of the Press Secretary, "Presidential Memorandum—Establishment of the Cyber Threat Intelligence Integration Center," February 25, 2015.

229 **"Such expansive development":** Charlie Savage and Jonathan Weisman, "N.S.A. Collection of Bulk Call Data Is Ruled Illegal," *New York Times,* May 7, 2015.

229 **wrote Judge Gerard E. Lynch:** Jonathan Stempel, "NSA's phone spying program ruled illegal by appeals court," *Reuters,* May 7, 2015.

232 **The president now has the authority:** The amended Robert T. Stafford Disaster Relief and Emergency Act of 1988.

EPILOGUE

238 **In his magisterial volume:** Richard Overy, *The Bombing War: Europe 1939–1945* (London: Allen Lane, 2013).

238 **infamously bombed the town of Guernica:** "The Bombing of Guernica, 1937," EyeWitness to History, 2005.

245 **A contemporaneous account in a local newspaper:** "Deposit Became Civil Defense Evacuation Zone in 1957 test," *The Daily Star,* November 12, 2012.

246 **In 1956 Congressman Chester E. Holifield:** *Civil Defense and Homeland Security: A Short History of National Preparedness Efforts* (Washington, DC: Department of Homeland Security, 2006).

247 **In July 1957 mock atomic bombs were dropped:** www .nebraskastudies.org.

247 **a bizarre exercise that involved a two-week survival test:** "Sheltering Cattle from Atomic Radiation in Nebraska," www .nebraskastudies.org.

ACKNOWLEDGMENTS

....

To acknowledge comes dangerously close to being obliged to admit something. It does not convey a sense of wholehearted enthusiasm. There is, in fact, something almost grudging about the word, and since custom dictates that an author acknowledge his indebtedness to those who contributed to the publication of a book, custom further undermines sincerity.

Perhaps if I explained how this book came to be . . .

I read a great deal and can't recall what, precisely, planted the idea of a cyberattack on the grid, but it seemed plausible. I was at home, looking out the window of an enclosed porch, wondering how I would care for my wife and myself if the power went out and stayed out. How would we survive? There are deer and wild turkey on our property and in the winter they roam the fields looking for food. I could shoot a deer, but I don't have a gun. I could buy a gun, but my chances of bringing down a deer would remain slim. Even if I got lucky, how would I skin and gut the animal? How would I preserve the meat? In very short order, I came to the conclusion that disaster survival is not my strong suit. Nor, for that matter, had I spent any part of my career investigating the survivability of the electric power grid.

I have, however, always been much taken by Will Rogers's observation that "we're all ignorant, just about different things." One of the great joys of a lifetime in my chosen profession has been my understanding that it entitles me to pursue and occasionally even harass the most knowledgeable experts in any field, and that they are obliged by some unwritten compact to answer questions from me, for no other reason than that I am a reporter.

For well over fifty years now, I have been enabled in this conceit by the woman I love: Grace Anne, the aforementioned wife. Her love for me has never wavered. She abandoned a brilliant academic career near the end of her work toward a PhD at Stanford to accommodate my youthful insecurities. Somehow she completed her studies as a Georgetown Law student while raising four children and catering to an ambitious husband. Scan a history of the last fifty years—the Kennedy assassinations, the Viet Nam war, the civil rights struggle, Nixon in China, troubles in the Middle East, Bosnia, South Africa, a succession of presidential election campaigns, two wars in Iraq—and you will have an incomplete index of why I was so often away from home and why Grace Anne's legal training was put to the service of managing her husband's production company and the family's financial affairs. That glosses over, entirely, the twenty-six years I spent anchoring *Nightline*. In such a context, it does not seem adequate to merely "acknowledge" her support during a further eighteen months of researching and writing a book. I can only embrace Grace Anne's latest round of sacrifice as yet further evidence of her great love and *acknowledge*, with a full heart, that love as the most precious gift I have ever received.

After a professional lifetime of working with a team of other reporters, producers, editors, and researchers, where the inter-action is constant, writing a serious book can be disorienting. There are many comparable interactions, but they tend to be conducted at arm's length and only sporadically. Writing, every-body understands, is lonely work. For all that, those who con-tribute to the endeavor are particularly important, in that they support morale even as they perform their own critical roles.

It does not take much to crush an idea. That's all a book is in its earliest stage. A dose of skepticism or ridicule from the right person at the wrong time can so undermine a writer's confidence that many wonderful ideas have surely been aban-doned because there was no Jonathan Segal to nurture them through their most fragile infancy. Jonathan is a brilliant editor at Knopf. We had worked together on a previous book of mine and hoped to collaborate on this one. That hope collapsed under the weight of my expectations, but not before Jonathan helped me shape the concept into something that captured the atten-tion of Bob Barnett. Bob's range as an attorney and advisor is frequently reduced to the phrase "power broker," or simply a recitation of a few dozen of his most famous clients. Among his clients, it is true, are those who have been or aspire to be presi-dent of the United States. I believe the current White House occupant has also engaged Bob's services at one point or another. I can only hope that he treats all of them with the same encour-agement, attention, and kindness that I have enjoyed. Engaging a publisher's attention is difficult enough. Convincing several of them to consider investing generously in the mere concept of a book requires more than I could have brought to bear on my own. Bob is not just a fine lawyer and a skilled negotiator; he has also been a good friend.

ACKNOWLEDGMENTS

Molly Stern is the publisher of Crown Publishers. She has the instincts of a good psychoanalyst and the nerve of a riverboat gambler. I have always had a particularly warm place in my heart for those who take a chance on me at a time when others are still expressing their reservations. That aura of confidence has persisted throughout the project. There is nothing that so lifts the spirits of a lagging writer as an ebullient message from the publisher. Molly does ebullient well. As does Crown's executive editor, Rachel Klayman, whose enthusiasm for this project, from its earlier stages on, never faltered.

A great deal of time has been wasted speculating on the possibility that an infinite number of chimpanzees banging away indefinitely on an infinite number of typewriters (it's an old idea) would eventually come up with the works of Shakespeare. To which I can only add: if they had Meghan Houser as an editor, their chances would be greatly enhanced. Her keen eye for organization often substituted clarity for confusion. To those who remain confused by one part or another of my book I can only explain that I did not always accept Meghan's recommendations. Meghan is a wonderful editor who never faltered in the face of difficult deadlines and an occasionally petulant author. She is at an early stage in her career, but before long, many will recognize her for what she is already—a great editor.

I have already referenced Ryan Ellis a couple of times in the body or in the endnotes of this book. He was brought to my attention by an old friend and colleague, David Sanger of the *New York Times*. David suggested that Ryan's work as a postdoctoral fellow at Harvard gave him the expertise to make him a reliable advisor on the vulnerabilities of the grid and the complex relationship between NERC and FERC. Ryan did, in fact, provide invaluable observations on early drafts of several chapters and on transcripts of numerous interviews. David Sanger, in turn, was most generous in introducing me to a number of very helpful sources. My old friend, Harry Rhoades, who, in my view, runs the classiest lecture agency in North America, extended himself repeatedly establishing connections for me with sources who were enormously important to this project. He did this, I should note, without ever violating their privacy but by encouraging them to contact me.

I suspect that almost every writer, at one point or another in a project, needs the support of an old friend. In the person of Tom Bettag, I also had a valued colleague. Together, over the course of almost two decades, we had collaborated on more than a thousand hours of television news programming. Tom knows how to most productively calibrate criticism and praise. He read an early draft, proposed some genuinely helpful changes, and, most important of all, provided encouragement at a time when it was sorely needed.

I had so many qualified applicants for the job of research assistant that I ended up hiring three. Rachel Baye, Katie Paul, and Morrow Willis were graduate students already working full-time jobs. We agreed that if each was occasionally available for this project, they would add up to the equivalent of a full-time research assistant. I don't know how Rachel and Katie found time for any life outside study and work. I can only say that neither ever let me down and both contributed mightily to the book.

I have saved word of Morrow to the last. Shortly after accepting my offer of a part-time job, Morrow learned that he had cancer and would have to return to his home in Texas for radiation treatments. I suggested that however much work he could do might be a distraction. It was, I said, of course up to him. Morrow eagerly took up the challenge and never missed a deadline. His research was clear, professional, and enormously helpful. Somehow, he managed to continue typing transcripts until the week before he died.

About three hundred of Morrow's friends and classmates gathered in a Georgetown University courtyard on a frigid winter evening to exchange memories of this remarkable young man. I had come to know his humor and his courage, but it was only through the reminiscences of his fellow students that I finally gained a more complete understanding of the man. Morrow and I talked often but we never met. It was my loss.

INDEX

....

INDEX

INDEX

INDEX

Ukraine, Russia and, 72–74
Underground, British, 242
unique exposure, 65
Unit 8200 (Israeli), 39
urban legends, 153–54
Utah Data Center, 71
Utah War (1857), 200

VE Day, 243–44
Vergeltungswaffen, 243
Verizon/Secret Service study, 42–43
vertical integration, limited
    competition vs., 27–28, 35–36
Viet Nam war, 57
Vinson, Amber, 41–42

Wall Street, 74
Wall Street Journal, 18–19, 22, 84–85,
    155
wards (Mormon), 189–90, 192, 203
"war games," 126
Washington County Education
    Center, 148
waste disposal, in disaster relief, 118,
    127, 158, 170, 234, 242
"water bomb," 154
water supply, 175, 225
    in disaster relief, 107, 118, 127, 136,

143–46, 148, 149, 154, 169, 173,
    187, 188, 207–8, 234, 242
weapons, in disaster relief, 144, 145,
    146, 149, 154, 156–60, 170–74,
    200, 204
    see also guns
Wellinghoff, Jon, 19
"whiteboarding," 48
Whitney Western Art Museum, 166
Williams, Brett, 45, 46–47, 86–87
wind power, 149, 155, 162
Wohlstetter, Albert, 88
Wolz, Stanley, 173–74, 177, 199, 205
Woolsey, James, 15, 22–24
World War I, 239
World War II, 162, 166–68, 225,
    237–44
Wyden, Ron, 218
Wyoming, western culture of,
    165–78

YouGov, 12
Young, Brigham, 184–85

Zetter, Kim, 88
Zevlin, Larry, 219
Zion, Mo., 182

ABOUT THE AUTHOR

TED KOPPEL was anchor and managing editor of ABC News *Nightline* from 1980 to 2005. Over those years he hosted more than six thousand programs, becoming the longest-serving network news anchor in U.S. broadcast history. Overall Koppel spent forty-two years at ABC News, serving as bureau chief in Miami and Hong Kong, covering events as diverse as Dr. Martin Luther King Jr.'s march from Selma to Montgomery and more than two years as a war correspondent in Viet Nam. In 2003 Koppel did his final stint as a combat correspondent, embedded with the 3rd Infantry Division during the invasion of Iraq.

As ABC's senior diplomatic correspondent, Koppel accompanied President Nixon on his breakthrough trip to China in 1972 and Henry Kissinger during his shuttle diplomacy in the Middle East. He covered every presidential campaign from Barry Goldwater in 1964 to Barack Obama in 2008. In 2012 New York University named Koppel one of the top 100 American journalists of the past 100 years. He has won every significant television award, including eight George Foster Peabody Awards, eleven Overseas Press Club Awards (one more than the previous record holder, Edward R. Murrow), twelve duPont-Columbia Awards, and forty-two Emmys. Since 2005 he has served as managing editor of the Discovery Channel, as a news analyst for BBC America, and as a special correspondent for Rock Center, and he continues to function as commentator and nonfiction book critic at NPR. He has been a contributing columnist to the *New York Times,* the *Washington Post,* and the *Wall Street Journal,* and is the author the *New York Times* bestseller *Off Camera.*